Principles of

AIR

QUALITY

MANAGEMENT

ROGER D. GRIFFIN

LEWIS PUBLISHERS
Boca Raton   Ann Arbor   London   Tokyo

Library of Congress Cataloging in Publication Data

Griffin, Roger D.
    Principles of air quality management / by Roger D. Griffin.
        p.   cm.
    Includes bibliographical references and index.
    1. Air quality management.   I. Title.
    TD883.G78   1994
    363.73'92--dc20   94-3857
    ISBN 0-87371-315-X (acid-free)

© 1994 by CRC Press, Inc.

Lewis Publishers is an imprint of CRC Press

No claim to original U.S. Government works

International Standard Book Number 0-87371-315-X

Library of Congress Card Number 94-3857

Printed in the United States of America
1 2 3 4 5 6 7 8 9 0

Printed on acid-free paper

DEDICATION

to those who seek the Truth in all things,
and to Him Who is

Roger D. Griffin received his BS from California State University at Long Beach and his MS from the University of California, Irvine in 1969 and 1976, respectively.

Mr. Griffin is a Registered Professional Engineer in Chemical Engineering in California. Additionally, he is a Diplomate in Air Pollution Control with the American Academy of Environmental Engineers and a Certified Hazardous Materials Manager, at the Master level, with the Institute of Hazardous Materials Management.

From 1969 to 1977, Mr. Griffin worked with local government agencies including the South Coast Air Quality Management District and its predecessor agencies. He served in various capacities beginning as a lab chemist analyzing well water, soils, and air samples for trace pollutants; as a field inspector; a source testing specialist; calibrating air monitoring instrumentation; running the emergency air pollution episode program; and as a permit processing engineer.

For the past 18 years, Mr. Griffin has been a consultant in the air quality management and hazardous materials fields. He has worked for private industry and government agencies in technology assessments for emissions control, air quality dispersion modeling, regulatory impact analyses, permitting assistance, hazardous waste incineration, process evaluations, solid waste-to-energy projects, chemical and low-level radioactive waste treatment and disposal, Superfund site remedial investigations and feasibility studies, assessment and remediation of property contamination for real estate transactions.

Mr. Griffin has presented papers and published articles in the United States, Europe, and Latin America on NO_2 emissions from methanol fired gas turbines, low-level radioactive waste incinerators, dioxin formation from MSW combustion, source reduction techniques, field testing methods for PCBs, and air emissions from industrial operations.

He is a member of the Air and Waste Management Association and the UCLA Advisory Board for the certificate programs in Air Quality and Hazardous Materials Management. He has taught the core courses in both the fundamentals of air pollution and the basic principles of hazardous materials management since the inception of the program.

ACKNOWLEDGMENTS

The following individuals are acknowledged for their contributions to this book and to a greater understanding of the field of air quality management:

Ms. Coralie Cupfer, Esq., formerly of the South Coast Air Quality Management District
Dr. Margaret Hamilton of the University of California, Los Angeles
Dr. Kathryn Kelly of Environmental Toxicology International
Dr. James Pitts of the University of California, Riverside
Dr. Stu Salot of CTL Environmental Services, Inc.
Dr. Larry Vardiman of the Institute of Creation Research

A special acknowledgment is given to my wife, Avice Marie Griffin, without whose encouragement this book would not have been possible.

PREFACE

In order to understand and manage our air quality resources, it is necessary to gain a fundamental understanding of the principles that govern our ability to do so.

From a local perspective, it may be considered desirable to install huge fans in order to "blow the smog away" from an area of high air pollutant concentrations, but from a technological and scientific perspective it is not feasible. Likewise, from a regional or continental perspective, it is not acceptable to merely transfer air contaminants from one location to another one by dilution or "blowing it away."

It is therefore the purpose of this book to give the reader a firm grasp of the principles that make up the broad field of air quality, its pollution, and its management. Starting from the basic definitions of air and types of air pollution we will follow some of its history through the present century. From that perspective, we will look at the terms used: air quality, emissions, standards and classifications of pollutants, and the production of secondary air pollution or photochemical smog. We next look at the health effects of the criteria air pollutants and those that are considered toxic or hazardous, and the effects of those contaminants on the human body. Air pollutant damages to materials and vegetation are also reviewed. The standards of acceptable air quality from the perspective of health impacts (chronic through emergency episode concentrations) and the techniques for measuring air quality are also reviewed.

We approach the sources of air contaminants from an anthropogenic as well as geogenic and biogenic perspective. Between sources and receptors we look at how contaminants are dispersed into the atmosphere from a local, regional, and global perspective. From these studies come an evaluation of the different models used to calculate dispersion and the models used to predict ambient air quality.

Federal laws and regulations as well as regional perspectives are summarized and evaluated. Control technologies that are available for both stationary sources and mobile sources are reviewed. From these, we are able to evaluate the possible management options for limiting emissions and optimizing air pollutant strategies.

Global air quality concerns, relative global emissions and the alternative views are evaluated from the perspective of management options that may be available to society at large. Of particular concern are those that may influence long-term air quality and health. Finally, we will be looking at indoor air quality and the future trends in air quality management approaches, with their limitations.

CONTENTS

Principles of

AIR

QUALITY

MANAGEMENT

1 THE ATMOSPHERE AND ITS CONTAMINANTS

Fair is foul and foul is fair, hover through the fog and filthy air.
Shakespeare, *Macbeth*

Air: composed primarily of two simple gases, oxygen and nitrogen, it is the one substance without which people cannot survive for more than about three minutes. Occupying less than 1/1,000,000 of the mass of the earth, the atmosphere is critical to life as we know it. "Pure air" is not an easily defined term, however.

While we may note the average concentrations of gases and other materials which make up the atmosphere, we are called upon to make a distinction between the air we breathe, termed the ambient air, and those trace constituents which may be considered pollutants or contaminants. *Contaminants* may be considered any materials other than the permanent gases seen in Table 1.1. A *pollutant,* on the other hand, has the connotation of being derived from mankind's activity. Again, this is an artificial distinction since gases, dust, and particles may be generated by natural processes (e.g., volcanoes) as well as by grinding operations which make a useful product such as cement. The health, visibility, and materials or vegetation effects may be the same.

In general, we are mostly concerned with air *pollutants* even though it is, as defined above, important to recognize that contaminants may or may not have adverse health effects. In one sense, windblown dust may be both a contaminant and pollutant. The distinction is further blurred when air quality regulations are adopted which are generic, i.e., based upon concentration measurements regardless of source. Thus, air contaminants may be

1

Table 1.1. Composition of the Clean Atmosphere Near Sea Level

Constituent	Chemical Formula	Percent by Volume	Parts Per Million by Volume
PERMANENT GASES			
Nitrogen	N_2	78.084	
Oxygen	O_2	20.946	
Argon	Ar	0.934	
Neon	Ne		18.2
Helium	He		5.2
Krypton	Kr		1.1
Hydrogen	H_2		0.5
Nitrous oxide	N_2O		0.3
Xenon	Xe		0.09
VARIABLE GASES			
Water vapor	H_2O	0.01–7	
Carbon dioxide	CO_2	0.035	
Methane	CH_4		1.5
Carbon monoxide	CO		0.1
Ozone	O_3		0.02
Ammonia	NH_3		0.01
Nitrogen dioxide	NO_2		0.001
Sulfur dioxide	SO_2		0.0002
Hydrogen sulfide	H_2S		0.0002

confused with air pollutants. Pollutants are a legitimate concern of society in dealing with air quality and its management, whereas contaminants may be accepted as a part of the natural world in which we live.

History of Air Pollution

Air pollution has been around as long as man has walked the earth. Indeed, in the earliest extant writings from the dawn of civilization, air emissions from forging operations for bronze, iron, and other implements were well known (*Genesis*, Chapter 4) and the emissions from smoking fires and blazing torches were also known (*Genesis*, Chapter 15). Natural emissions from volcanoes and forest fires were well known and a part of everyday life. The apparent response was to let nature take its course and blow air contaminants away.

It has been known for millennia that people with certain occupations such as miners incurred diseases of the lungs and respiratory system. Looking back today we know that many of these effects were not only induced by fine particles and toxic metals, but also carbon monoxide. It appears that radon gas was one of the components of miners' lung disease from the 15th through the 20th century in the Erzgebirge mountains in central and eastern Germany.

Emissions from cooking, heating, and fires in general were well known. Apart from natural causes, the contaminant sources were simple solid fuels and metal working operations. Odors were well known as a part of common life. Attempts to stay out of "bad air" *(mal aria)* or areas where refuse, garbage, and human corpses were deposited in ancient times was the common approach to air quality management.

Maimonides of Egypt in the 12th century documented the effects of polluted air on child mortality and morbidity in the Cairo of his day.

In England in 1228, coal smoke was determined to be "detrimental to human health" since Queen Eleanor of Aquitaine had complained that it hurt her lungs. She left the palace in Nottingham in favor of the less polluted countryside. The first known attempt to "manage" air quality in England was the prohibition on burning soft coal in 1273 A.D. By the end of the 13th century, there was considerable agitation about the use of sea-coal, and complaints were mounting. In 1306 King Edward I attempted to drastically curtail the use of coal in London by passing a law stating that "no coal was to be burned by industry and artisans during Parliament." "Fumifugium" was a treatise by John Evelyn published in 1684 which dealt with smoke, its deleterious qualities, and offered some remedies. The typical response was to build chimneys as urbanization and population centers grew. These provided some relief in the immediate vicinity, but contributed to an overall deterioration of air quality in the surrounding areas.

With the rise of industrialization in the late 18th and 19th centuries, the effects of air pollutant emissions were noted on greater portions of the population; however, air quality management options, i.e., controls and prohibitions, generally lagged behind societal concerns for sanitation, water supply, and solid waste. It is noted that odors from the Thames River in the late 1850s were reported to make life in London almost intolerable. Industrial centers such as the Ruhr Valley in Germany were known to significantly impact life in general, and health and appearance in particular.

The effects of air pollution on vegetation in the U.S., primarily sulfur oxides and particulate matter containing heavy metals, were seen notably in Tennessee near Ducktown in the Copper Hill area (the hometown of the author's father). Prior to 1864, the area was covered with pine forests and shrubs, but following the start-up of the copper ore smelter, vast quantities of air pollutants were emitted directly into the air without controls. The surrounding countryside was totally denuded. This included 7,000 acres of forest and over 17,000 acres of land which could no longer support vegetation other than a few grasses.

The 20th Century

It wasn't until the 20th century that air quality management approaches (i.e., air pollution control regulations, primarily smoke and odor abatement measures) began to make their appearance. In 1906, Frederick G. Cottrell invented the first practical air pollution control device: an electrostatic precipitator to control emissions of acid droplets from a sulfuric acid manufacturing plant.

The term *smog*, a contraction of the words smoke and fog, was popularized in Great Britain as a result of a report by a Dr. H.A. Des Voeux, a London physician who worked in the field of public health, to the Manchester Conference of the Smoke Abatement League in 1911. It concerned conditions in Scotland, where the combination of smoke, sulfurous gases, and fog were believed to have claimed over 1,000 lives in Glasgow and Edinburgh in 1909.

Man-made air emissions were generated in that period from two general sources: industrial operations and generation of electrical power. Most of these operations were driven by the combustion of solid fuels. Air emissions were without controls and were directly emitted to the atmosphere through a chimney or stack. The early control efforts were aimed at either recovery of materials (i.e., acid) emitted from the plant, or abating smoke emissions.

During this era, efforts were made at understanding not only emissions but also their dispersion. The earliest attempts to understand the dispersion of air contaminants were made by observers and meteorologists of the British army artillery corps during World War I and afterward by observations of antiaircraft shell bursts.

As industrialization proceeded in the early decades of the 20th century, significant air pollutant "episodes" were noted, such as the condition which occurred in the Meuse Valley in Belgium in 1930. In this situation, meteorological conditions contributed to air stagnation which led to a buildup of pollutant concentrations from a number of industrial sites in the vicinity. In this incident approximately 60 people died and a large number of others became ill.

It wasn't until the post-World War II era, however, that a better understanding of air pollution was gained and control technology approaches were instituted. In the United States in the 1940s, in addition to efforts by pioneers to understand air pollutants, changes were also made in the fuels used (i.e., from coal to oil and gas). More significantly, the number and types of sources changed dramatically. The latter refers to the dramatic increase in the number of automobiles in the United States in the post-World War II era.

In the late 1940s and early 1950s in Los Angeles, a new type of air pollution was noted, *photochemical smog*. We now know this type of air pollution is made up of photochemical oxidant gases. Since that time, photochemical air pollution has been found in virtually every urbanized area in the world. It results from a combination of weather patterns, sunlight, and specific emissions interacting to form ground-level ozone.

TERMS AND DEFINITIONS

In order to clarify our understanding it is necessary to take a look at terms which are in use throughout this field.

Ambient Air

The ambient air refers to that portion of the atmosphere which is in the "breathing zone" of the inhabitants of the earth's surface. As such, it is limited to the lower several hundred feet of the earth's atmosphere. The ambient air and, in particular, those contaminants which may exist in the ambient air, are of major concern since they determine the effects on human health, materials, and vegetation. The concentrations of contaminants in the atmosphere are the implied subject when we speak of ambient air and ambient air quality.

Criteria and Noncriteria Air Pollutants

There are two basic categories of contaminants in the ambient air which are of concern. These two categories are the *criteria* pollutants and the *noncriteria* pollutants.

The **criteria** air pollutants are those air contaminants for which federal concentration limits have been set as the dividing line between acceptable air quality and poor or unhealthy air quality.

The ambient air quality (AAQ) standard is that *concentration* of a given air pollutant in the ambient air over a specified period of time below which the U.S. Environmental Protection Agency (EPA) believes there are no long-term adverse health effects. The criteria air pollutants include four gases and two solids:

- nitrogen dioxide (NO$_2$)
- sulfur dioxide (SO$_2$)
- carbon monoxide (CO)
- ozone (O$_3$)
- particulate matter (PM$_{10}$), and
- lead (Pb)

The criteria pollutants have been studied in some cases for over 100 years and their human health and vegetation effects are fairly well documented.

The *noncriteria* pollutants are those other contaminants designated as toxic or hazardous by legislation or regulation. They fall into two further categories depending upon the legislation which defines them. In general, the hazardous air pollutants (HAPs) may pose a variety of health effects (irritation, asphyxia, etc.), whereas the "toxics" focus on one physiological response (i.e., toxicity).

These noncriteria air pollutants have been studied in industrial hygiene settings. In the ambient air, noncriteria pollutants tend to be several orders of magnitude lower in concentration than the criteria pollutants. For instance, it would not be uncommon to find ambient carbon monoxide in the parts per million range whereas ambient concentrations of a hazardous air pollutant, such as benzene, would be in the parts per billion range.

The concentrations of ambient air pollutants are expressed in terms of concentration: either in terms of a mass per unit volume such as $\mu g/m^3$ (micrograms per cubic meter), or in terms of a pure volumetric ratio: volumes of contaminant per million volumes of air. The relationship between the two concentration expressions is a function of the molecular weight of the contaminant in question and the standard conditions referenced in the setting of the ambient air quality standards. The conversion between mass units and volumetric ratios at standard temperature and pressure is:

$$\mu g/m^3 = ppm \times MW \div .02445 = ppb \times MW \div 24.45 \quad (1.1)$$

where: $\mu g/m^3$ = micrograms per cubic meter
  ppm = parts per million by volume
  MW = molecular weight of the contaminant
  ppb = parts per billion

It should be noted that *ambient air quality standards* have differing time periods over which the concentration is to be calculated. In some cases (ozone, for example) these are often expressed as parts per hundred million, or pphm.

Emissions

Emissions are air contaminant mass releases to the atmosphere from a source. They may be from a tail pipe, vent, or stack, though some may be airborne, such as those from aircraft. Emission *regulations* are expressed in one of two ways: the first is by a mass emission *rate*, such as pounds per hour, tons per year, or milligrams per second. Or they may be expressed in terms of a *concentration*, such as parts per million or grains per standard dry cubic foot (gr/SDCF).

Standards for either expression are set by regulation, depending upon the air contaminant, the source, and the regulatory jurisdiction. Mass emission rates are calculated from a measured concentration and calculations of total gas flow per unit time. Emission standards expressed in concentration units were derived from early measurements of air pollutants at sources. In these early efforts to control air pollution, this was the quickest method to determine compliance at a source without having to quantify the mass emission rates.

The Epidemiologic Model

Mass emission rates and ambient air quality concentrations are related by the source/transport/receptor model. In the public health field, this is termed *the epidemiologic model*, and relates a source through a mode of transmission to the receptor.

Modeling of the transport phenomenon during which diffusion and dispersion occur yields a calculated downwind ambient pollutant gas concentration. Recent air quality management approaches for hazardous air pollutants and air toxics have used the epidemiologic model to relate health effects to emissions by using various dispersion models.

COMPONENTS OF THE ATMOSPHERE

Fundamental to understanding air quality management is a basic understanding of the atmosphere itself: its composition, functions, movements, and impacts.

The major components of the atmosphere on a dry basis, as seen earlier, are: nitrogen, at about 78% and oxygen, at about 21%; leaving about 1% for all the other components. Neither nitrogen nor oxygen are passive; they have significant impacts of their own, particularly when they react with each other to form other air contaminants. The gas concentrations seen in Table 1.1 vary somewhat from point to point across the surface of the globe.

The major components in the atmosphere function first in bulk transport. This includes the transport of energy from east to west and pole to equator, as well as the transport of contamination from a source to a receptor. Finally, they participate in the nitrogen cycle which is critical for the functions of photosynthesis and crop production. Atmospheric temperature and pressure are functions of these bulk gases.

The minor components of the atmosphere include the gases water vapor (H_2O) and carbon dioxide (CO_2). These components vary much more in concentration; several percent in the case of water, and several parts per million for carbon dioxide. These two gases have major impacts in and of themselves.

First, water participates in the hydrologic cycle on which life itself depends. In addition, water in the vapor or gaseous state is the most important gas contributing to the earth maintaining a livable temperature (the greenhouse effect). It is a unique substance because at the earth's surface it exists in all three material states: solid, liquid, and gas. Water performs its major functions in the life processes as a liquid over a narrow temperature range of only 100° Kelvin in a universe which tends to operate either near absolute zero or at millions of degrees.

The other unique properties of water are the large amounts of energy per unit mass that are either absorbed or released when water moves between solid and liquid, or between liquid and gaseous states. These have tremendous impacts on the energy balance in the atmosphere, and are responsible for the "weather." Carbon dioxide contributes to the greenhouse effect, but at a much lower contribution level. However, CO_2 participates in the carbon cycle (photosynthesis and respiration), without which life also could not exist.

The final materials which exist in the atmosphere at highly variable concentrations are the trace components. These include the carbon gases: carbon monoxide (CO) and methane (CH_4); the nitrogen gases: ammonia (NH_3), oxides of nitrogen (NO_X = NO and NO_2), and nitrous oxide (N_2O); the sulfur gases: hydrogen sulfide (H_2S) and sulfur dioxide (SO_2); ozone (O_3); and particulate matter (finely divided solid or liquid particles suspended in the atmosphere by mechanical mixing or Brownian motion).

These trace components appear to be closely intertwined, both in terms of their influence and their reactions with each other in photochemistry. For instance, methane (CH_4) is not an inert gas, but appears to participate extensively in reactions with carbon monoxide, oxygen, and hydroxyl radicals to form ozone.

These trace contaminants also have major influences on health, materials damage, and impacts on vegetation and crops.

Physical Characteristics

Air is a mixture of all of these gases within an approximate molecular weight of 29. As a gas, it follows the physical gas laws. Within a somewhat restricted operating temperature (ambient conditions: typically 0 to 50°C), it follows the ideal gas law:

$$PV = nRT \qquad (1.2)$$

where:
P = the absolute pressure
V = the volume of the gas parcel in question
n = the number of moles of gas in the parcel
R = the gas law constant, and
T = the absolute temperature

Air is a fluid and therefore tends to move under the influence of external forces as any fluid such as water would. However, because it is a gas, it has no effective shear strength and will change its volume and density as the pressure and temperature change.

Each component in the atmosphere has its own partial pressure which is independent of all other gas components. Because of this, the molecular weight of the atmosphere changes as one approaches the stratosphere.

Standard Conditions

Because the atmosphere varies from place to place in terms of temperature (T), density (ρ), and pressure (P), a *reference pressure* has been defined as one "standard atmosphere at sea level." Sometimes the standard atmosphere is considered under "dry" conditions. In these cases, all water vapor is subtracted and the remaining gas constituents are calculated on a "dry basis." The equivalencies between pressure measurement units at one standard atmosphere are:

$$1,010 \text{ mb} = 760 \text{ mm of Hg} = 14.7 \text{ psi} \qquad (1.3)$$

where: mb = millibar
         mmHg = millimeters of mercury column, and
           psi = pounds per square inch

Standard temperatures tend to vary, depending upon the reference. For the U.S. EPA, the reference temperature is 20°C. For other air quality jurisdictions the standard temperature is 60°F. In other fields the standard temperature may be 32°F or 273°K.

Density is the mass to volume ratio of any given parcel of air. The correction of a given density to a standard condition is a straight proportionality between the ambient P and T and the reference density at standard T and P. For 14.7 psi and 60°F, the dry air density is 0.0763 lb/cubic foot. At 760 mmHg and 20°C, the dry air density is 1.356 grams/liter.

Dew Point and Humidity

As the temperature of a gas parcel increases, its capacity to hold water vapor increases. The measure of the amount of moisture or water vapor in a given parcel of air is called its *absolute humidity* (i.e., milligrams of moisture per cubic meter of air). If one compares the actual amount of moisture in an air parcel to the maximum which it could hold at a given temperature, we have *relative humidity*. It is expressed in percent for the given temperature and atmosphere pressure being considered. If the parcel of air being considered contains all of the possible water that it could theoretically hold, we term that condition as *saturation* or 100% relative humidity.

Because saturation is a function of temperature, if we have a parcel of air which is already saturated at some temperature (such as 90°F), and that parcel of air is cooled, the water vapor will condense out to form droplets of liquid water. These will be in the form of a fog. Depending upon the other constituents present in that parcel of air such as particles, larger droplets will precipitate out. To condense water from gas to liquid, a large amount of heat ("the heat of condensation") must be removed. This latter characteristic has significant effects on the atmosphere.

STATES OF AIR POLLUTANTS

In understanding the states of air pollutants we realize that we are dealing with all three states of matter: solid, liquid, and gas. Each has its own characteristics and impacts.

Pollutant Gas Features

The major pollutant gases include sulfur dioxide, carbon monoxide, volatile organic compounds, nitric oxide, and nitrogen dioxide. Thus they obey the gas laws outlined earlier. In general, they are reactive — either by way of gaseous photochemical reactions or by direct reactions with materials, vegetation, or living tissues. There are significant size features: these gas molecules are on the order of 0.0005 micrometers (μ, or microns) in diameter. They mix in a closed environment by diffusion and in the atmosphere at large by convection.

There is little visibility impact from the true pollutant gases other than changes to the color of the atmosphere as we view it. For instance, NO_2 absorbs light in the UV end of the visible spectrum. Therefore, when NO_2 is present in moderate concentrations we see a yellow to brown coloration. Another salient feature is that these gases are primarily acids and thus they participate in all of the acid reactions common to the field of chemistry. The health effects of these gases are very important.

Finally, there is high annual emission rate (tonnage) in the U.S. for these anthropogenic emissions. In the U.S., about 120 million metric tons per year of gaseous air pollutants are emitted to the atmosphere. Gaseous emissions are well in excess of 90% of all U.S. air pollutant emissions. Over half of this total is carbon monoxide.

Particulate Features

Particles or particulate matter suspended in the atmosphere that are either solids or liquid droplets are termed aerosols. Particulate matter (PM) in solid form is not particularly reactive unless it encounters liquids (i.e., condensing water vapor or other liquids). Liquid droplets are considered particulate matter or *particulates* in some jurisdictions. The pH of the droplet as well as its composition and the amount of dissolved material in the droplet is of concern. Particulate matter represents approximately 6% of the total anthropogenic burden of emissions in the U.S. at approximately 7 million metric tons per year.

The size of particulates is highly variable, ranging from less than 0.01 μ to over 100 μ. In Figure 1.1 we see a typical ambient air *trimodal distribution* of particulate diameters vs number density. These three size groupings are generally attributable to different sources.

The largest particles, those with an aerodynamic diameter larger than about 10 microns, are generated either by natural causes (windblown dust) or by attrition or grinding of naturally occurring materials, such as sand, coal, lime, etc. The largest ones fall out soonest and settle out (under quiet conditions) with a constant velocity within a fairly short distance. They are governed by Stokes' law, which relates the settling velocity to the particle density and diameter. Figure 1.2 indicates the particulate size ranges of various materials.

At the other end of the distribution we find the fine particles in the 0.002 to 0.1 micron range. These are particles that behave like gas molecules and exhibit Brownian motion; thus they remain suspended in the atmosphere for long periods. This size range is termed the *condensation nuclei* range because they are formed by the condensation of gases or vapors into liquid or solid droplets. Sources of these particulates include combustion and fumes from metallurgical operations. The key feature of these particles is that: (1) they are very reactive on their surfaces, and (2) they tend *not* to

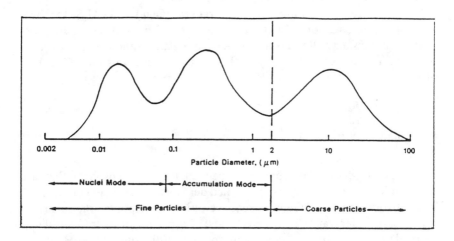

Figure 1.1. Modal distribution of atmospheric particles.

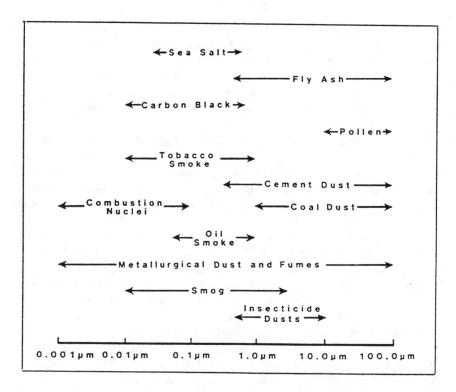

Figure 1.2.   Size ranges of common atmospheric particles.

remain in that particle size. Also, they serve as sites for the deposition of heavy metals, such as selenium, arsenic, and lead.

The middle distributions, roughly between 0.1 and 2 microns in diameter, form the *accumulation* group. These are termed "accumulation" since this size range is the final resting place of particles from the other two regions. Particles in the condensation nuclei range tend to agglomerate and grow to an optimum diameter between approximately 0.3 to 0.5 microns under the influence of Brownian motion and condensation processes. The growth tends to stop there, due to surface charge and surface tension effects. Coarse particles will continue to break down due to mechanical action and weathering. They tend to accumulate in the 1 to 2 micron diameter sizes. Particulates in the less than 2.5 μ diameter range are considered fine particles primarily due to their deposition sites within the lungs. Those larger than 2.5 μ are termed coarse particles.

It should be noted that these diameters are not the true physical diameters of the particles. Particles tend to be irregular in shape except for liquid droplets. *Aerodynamic diameter* is the diameter of the particle in question equivalent to a spherical particle of unit density (1 gram per cubic centimeter). Instruments are calibrated to the unit density spherical particle and therefore measure particulate size in terms of aerodynamic diameter. Optical measurements, of course, reveal a wide variety of shapes and aspect ratios.

Liquid aerosols are less well defined. Several considerations differentiate them from gases and solid particulates. These aerosols are typically classified in the 0.01 to 1.0μ aerodynamic diameter size range. Also, because aerosol diameters are at the wavelengths of visible light, optical scattering occurs. An aerosol's ability to interfere with light transmission is also related to its mass concentration.

The components of aerosols include soil-derived crustal materials; water; organic compounds; water-soluble acids such as HCl, SO_2, and carbon dioxide; ions derived from other, originally gaseous, pollutants such as ammonia; and finally, elements such as soot or black carbon in its elemental form, mixed with the other materials. As such, aerosols are heterogeneous mixtures and therefore their composition varies widely from location to location, time of the year, meteorological conditions, etc.

Aerosols are reactive and will participate in the formation of other compounds within the droplet itself. Figure 1.3 shows an idealized schematic of the composition of atmospheric aerosols and their sources.

Two of the greatest impacts of aerosols are on health and visibility. Due to the size of aerosols, they are deposited deep in the lungs (in the alveoli), where they accumulate. From there, contaminants are carried across tissue membranes into the body.

CONTAMINANT CLASSIFICATIONS

Air contaminants can be classified in one of two major categories, depending on the generation mode: primary emissions and secondary. The latter are produced by further reactions in the air or within liquid aerosols.

Primary Air Contaminants

Within the primary classification are two major subdivisions: those which are naturally occurring and those which are anthropogenic or derived from the activities of mankind. In the chapter on sources we will look at the

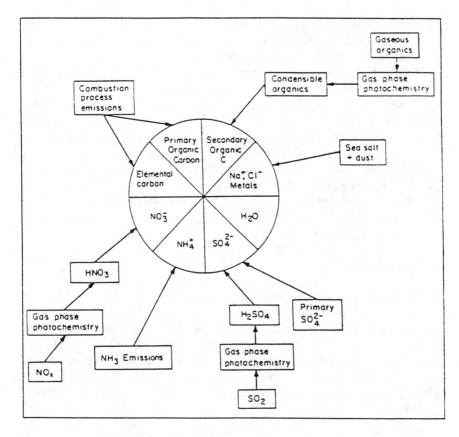

Figure 1.3.   Idealized schematic of the composition of atmospheric aerosols.

relative contributions of the two classifications. Within the naturally occurring category we find those which are geogenic or derived from actions from the earth's crust, and those which are biogenic or derived from living organisms.

Natural Emissions

The geogenic source contaminants include the sulfur gases (SO_2 and H_2S), carbon dioxide, particulates (minerals), chlorides and other ions such as sodium (from volcanoes and sea salt aerosols), and the petroleum gases.

The latter are derived from deposits within the earth's surface and contain all of the organic compounds normally found in crude oil systems, from methane to the heaviest asphaltic compounds.

The biogenic gases include hydrogen and all of the carbon based gases derived from biological activity. These processes include photosynthesis, metabolic action, decomposition of living matter, and emissions from plants and animals. Typical of these carbon gases are CO, CO_2, methane, CH_3Cl, and the terpenes (organic compounds whose basic structures are formed from isoprene—a five-carbon olefinic chain compound). The latter are the primary gases released by land-based plants. The terpenes are highly photochemically reactive, and present significant emissions which lead to secondary reactions in the atmosphere. Emission rates of those compounds are a strong function of temperature.

Anthropogenic Emissions

The anthropogenic pollutants (i.e., generated by man's activities) include CO and sulfur dioxide, hydrochloric acid, NO_x, as well as the halogenated solvents such as chlorofluorocarbons. In addition, particulates, covering all of the size ranges noted earlier, come from anthropogenic activities. Petroleum related gases are also included in the anthropogenic category, because they are also released to the atmosphere from mankind's activities in handling, processing, transporting, reacting, and combusting petroleum based materials.

Secondary Air Contaminants

Secondary air contaminants are those produced either by reactions in the gas phase or in aerosols suspended in the atmosphere. *Oxidants* are the most important class of secondary pollutants produced in the gas phase.

Also, particulates are produced from primary pollutant gases. For example, NO (from combustion operations or natural causes) is initially oxidized to NO_2, and ultimately to nitric acid and ionic nitrates. Likewise, sulfur dioxide is oxidized in the atmosphere to sulfur trioxide which, in the presence of moisture, forms sulfuric acid, and ultimately, particulate sulfates.

Within an aerosol, further reactions occur to change the chemical composition and potential health effects of the droplet. The acid forms of sulfur and nitrogen oxide gases will react with naturally occurring ammonia to form the salts: ammonium sulfate and nitrate.

PHOTOCHEMICAL SMOG

The early researchers were aware of the nature of photochemical oxidants but were unable to determine the relationships of the measured species with the activity of sunlight to form the oxidizing compounds. In Los Angeles, researchers in the 1950s discovered a fairly common pattern of several of the pollutant gases which were noted during the course of a typical "smoggy" day in Southern California (Figure 1.4).

The initial levels of contaminants remain fairly constant until about sunrise. During the morning hours sharp increases in NO and nonmethane hydrocarbon concentrations were noted. It was observed that this corresponded to the starting and driving of mobile sources of combustion contaminants, principally NO and hydrocarbons from gasoline powered automobiles.

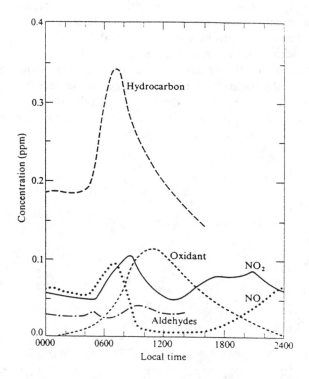

Figure 1.4.   Daily variation in air pollutant levels.

By late morning, the concentrations of hydrocarbons and NO began to fall at about the same time that NO_2 concentrations began to rise. This was followed at mid-day by declining concentrations of NO_2 with increasing concentrations of oxidants, principally ozone. As the afternoon proceeded and the sun dipped it was noted that oxidant concentrations also began falling off. This was followed by a late afternoon surge in NO and NO_2 readings, which dropped off and leveled out as the evening progressed.

These early researchers concluded that there was a relationship between sunlight, hydrocarbons/volatile organic compounds, NO and NO_2. This led other researchers to investigate the relationship between organic gases and oxides of nitrogen in smog chamber studies in the presence of sunlight to form oxidants or ozone.

Carbon monoxide concentrations (Figure 1.5) also showed a diurnal pattern with rises in the early morning and late afternoon periods. CO is not a direct participant in the majority of the reactions producing oxidants in the atmosphere. CO *competes* with the formation of ozone by reacting with OH free radicals to produce a free hydrogen atom. The latter reacts rapidly with oxygen to form the hydroperoxy free radical (HO_2), which may subsequently participate in the ozone formation.

Further investigations of atmospheric reactions led ultimately to the understanding that oxidants (primarily ozone) are formed by two mechanisms: one is a direct photolysis whereby NO_2 is split by blue/violet light to form NO and an oxygen atom free radical. The resulting oxygen atom

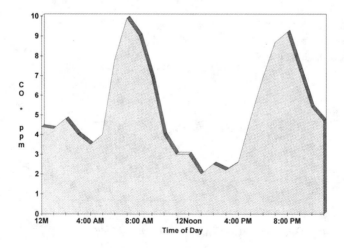

Figure 1.5.   Typical urban carbon monoxide levels.

quickly combines with molecular oxygen in a three-body reaction identical to the one important for the production of ozone in the stratosphere:

$$NO_2 + h\nu \Rightarrow NO + O^* \qquad (1.4)$$

$$O^* + O_2 + M \Rightarrow O_3 + M \qquad (1.5)$$

M represents any other molecule (typically N_2 or O_2) available to carry off excess energy. The two reaction components, NO and O_3, will react with each other to form $NO_2 + O_2$ once again. Thus an equilibrium is established under ideal conditions whereby ozone is formed by the action of sunlight on NO_2 in the atmosphere. Figure 1.6 shows this simple NO_2/O_3 cycle.

Calculations, however, for these equilibrium concentrations of ozone do not match the higher concentrations measured in the photochemical smog of Los Angeles. For example, for an initial concentration of 0.1 ppm NO_2 the predicted ozone concentration should be about 0.03 ppm, whereas the observed ambient concentration is approximately 10 times higher. Thus, this simple reaction set is not sufficient by itself to explain the formation of photochemical ozone.

Further studies indicated that the key to understanding elevated ground-level ozone levels are found in the chemical reactions which convert NO to NO_2 without consuming O_3. It was discovered that a critical component of the overall higher production of ozone was the impact of reactive hydrocar-

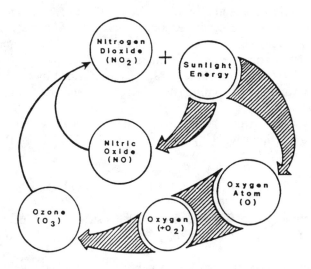

Figure 1.6. Photolysis of NO_2 and generation of O_3.

bons. They are generated by volatile organic compounds (VOCs) combining with OH* free radicals in the vapor phase. A simplified reaction sequence then becomes:

$$R - C - H + OH^* \Rightarrow R - C^* + H_2O \tag{1.6}$$

$$R - C^* + O_2 \Rightarrow RO_2^* \tag{1.7}$$

$$RO_2^* + NO \Rightarrow NO_2 + RO^* \tag{1.8}$$

then the original reactions occur:

$$NO_2 + h\nu \Rightarrow O^* + NO \tag{1.9}$$

$$O^* + O_2 + M \Rightarrow O_3 + M \tag{1.10}$$

where: h = ultraviolet radiation
R is a hydrocarbon fragment
X* is any free radical atom

The overall equation then becomes:

$$RO_2 + O_2 + h\nu \Rightarrow RO + O_3 \tag{1.11}$$

A schematic of this overall reaction is given in Figure 1.7. All of the species, such as OH, are produced by reactions involving photocatalyzed associations of O_3, aldehyde, or carbonyl compounds and HNO_2. As noted earlier, CO may also participate in the formation of HO_2 radicals by reaction of OH to form $H^* + CO_2$. The OH radicals are formed by reactions of O^* with water vapor or the photolysis of aldehydes and nitrous acid to yield OH* species.

The RO_2 species [from the original hydrocarbon chain with an acyl (-CO-) unit from a reaction with NO_2] forms the peroxyacyl nitrate, or PAN, family of compounds:

$$R\text{-}CO\text{-}O_2^* + NO_2 \Rightarrow R\text{-}CO\text{-}O\text{-}O\text{-}NO_2 \tag{1.12}$$

Typical compounds which give rise to such species would include volatile organic compounds such as acrolein ($CH_2 = CH\text{-}CO\text{-}H$) and formaldehyde.

In summary, the secondary photochemical pollutants derive their energy for breaking and forming new chemical bonds from ultraviolet radiation in

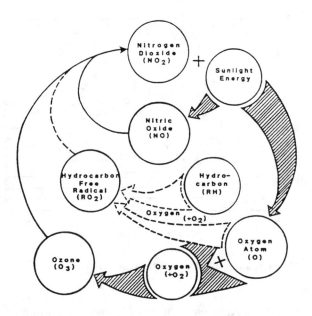

Figure 1.7. The influence of HCs and free radicals on atmospheric O_3 generation.

sunlight. Thus, if sunlight is absent, photochemical oxidants will not be formed. Oxidants include ozone (O_3) and PAN (peroxyacetyl nitrate). Minor quantities of NO_2 and H_2O_2 are formed in addition to the radicals OH, HO_2, and partially oxidized hydrocarbons.

Air Quality Management Aspects of Photochemical Reactions

Knowledge of these reactions provides us with the opportunities for implementing strategies for managing or improving air quality in locations subject to photochemical oxidants. The current state of knowledge indicates that both the NO_x and reactive organic gases are the key components.

Figure 1.8 is a computerized simulation (the EKMA model) of known chemical reactions from smog chamber studies where oxides of nitrogen and hydrocarbon species are irradiated with sunlight. From this simulation we have a contour map of maximum ozone production for differing initial concentrations of NO_x and hydrocarbons. This indicates that with high

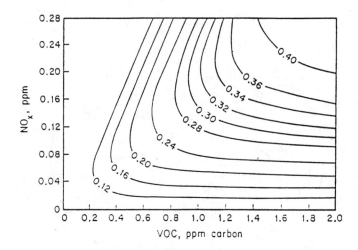

Figure 1.8.   Modeled ozone generation as a function of NO_x and VOC.

concentrations of both NO_x and VOC hydrocarbons, maximum ozone concentrations of 0.4 parts per million are possible. With a decreasing NO_x and hydrocarbon concentrations, maximum ozone concentrations will also decrease.

However, if both concentrations of contaminants are progressively lowered at the same *ratio* of hydrocarbons to NO_x, the maximum ozone concentrations may only slowly decrease or even increase! On the other hand, if control strategies are implemented such that the ratio of either one or the other contaminant decreases dramatically from some initial point, the minimum ozone concentration may be attained much more quickly.

It must be remembered that this is only a generic output of modeled atmospheric reactions — the *shapes of the actual curves are unique to each location*. These results, based upon evaluations of ambient concentrations and smog chamber studies, *point the way* toward potential air quality management strategies; the actual strategies may be different for each location.

We must also be aware that contaminants such as hydrocarbons are pollutants in their own right. They are thus subject to regulations and controls due to their health impact, as well as materials damage and effects on crops and vegetation. These are the subject of the next chapter.

2   EFFECTS OF AIR POLLUTION

All things are poisons, for there is nothing without poisonous qualities. It is only the dose that makes a thing poison.
Philippus Theophrastas Bombast von Hohenheim
(Paracelsus), ca. 1540

Air pollution is known to cause damage to the entire ecosystem. Such damage includes adverse effects on human health, damage to crops, vegetation, and forests, and finally, to materials that we all use.

Our first order priority is to review and evaluate effects on human health of the different air contaminants. Following that, we consider effects on vegetation, crops, and finally, materials.

SOURCES OF HEALTH EFFECTS INFORMATION

What we know about the effects of air pollution on people comes to us from three major sources. Each of these is needed in order to gain a good understanding of air pollution's effects. These three are: animal studies or evaluations, human exposure studies, and epidemiology.

Animal effect studies are used to a large extent because we can control the conditions surrounding the animals. We are also able to perform long-term (potentially generational) studies on test animals. This is a positive aspect since we are able to monitor and control conditions. Unfortunately, due to differences noted later, the health effects for a given exposure to an air pollutant are not necessarily transferrable in terms of its effects on human beings.

The second technique to gather effects data is by human exposure studies. These are controlled conditions in a laboratory using exposures of specific contaminants to real people; therefore, the results are directly transferrable to different segments of the population. In one sense this may provide us with the best "data" since it is transferrable, but such exposure studies are *limited* because we cannot knowingly expose individuals to cancer-causing agents or those which cause reproductive toxicity or other life-threatening contaminants.

These studies tend to be rather expensive and are generally limited to a maximum period of eight hours. Therefore, no chronic effects may be studied. Likewise, we are not able to gain an understanding of subtle, or microcellular, effects which may occur under human exposures to various air contaminants.

Epidemiology is another approach where we are able to study human responses to air contaminants. In many cases this may be done over long time periods. Thus the data may allow for evaluations of chronic exposures. Also, we are dealing with real people. The limitations are that the researchers have a very limited control over significant variables involved, such as lifestyle, smoking habits, age, sleep patterns, nutrition, etc. These studies tend to be costly since a great deal of time must be spent observing individuals and accessing medical records over a period potentially of years.

Sensitive Population Groups

In addition to concerns for sources of health information, we find that the general population ranges from vigorously hearty individuals to those who are particularly susceptible to ambient pollutants. Table 2.1 lists those general population categories which are considered sensitive to air pollution.

Within each of these categories there may be some overlap of effects as well as the possibility of multiple sensitivities. Pregnancy is another consideration. It has been estimated that roughly 7% of the population already

Table 2.1. Population Groups Sensitive to Air Pollution

Group	% of Population
Cardiovascular disease	7.
Chronic respiratory disease	8.9
Elderly (>65 years)	9.6
Children (<14 years)	20.
Athletic activities	7.

suffers from some forms of cardiovascular disease, with approximately 9% susceptible to chronic respiratory diseases. Ten percent may be considered elderly, with possibly 20% children under the age of 14. On the average, perhaps an additional 7% of the general population may be involved in some form of vigorous athletic activity. Again, while not exclusive, these numbers would indicate perhaps half the total population in any location may have existing sensitivity to air pollution.

CRITERIA VERSUS NONCRITERIA AIR POLLUTANTS

There are differences between how criteria pollutants act and how the noncriteria pollutants act. Prior to an evaluation of specific pollutant effects, it would be instructive to differentiate between these two types of air pollutants. Table 2.2 summarizes these differences. Lead, a criteria pollutant, is the exception to this discussion, since it is obviously toxic.

In general, criteria pollutants have a known *threshold* dose, below which no adverse health effects are known to remain after cessation of exposure. With respect to carcinogens, there is no known threshold to which we can point with confidence.

In comparison to the noncriteria or hazardous air pollutants (HAPs), which are potentially numerous, there are only six criteria air pollutants. Other significant differences are that the gaseous criteria pollutants occur in the ambient air at the parts per million level, whereas the others tend to be

Table 2.2. Criteria and Hazardous Air Pollutant Comparisons

Criteria Pollutants[a]	Hazardous Air Pollutants
Few (6)	Potentially numerous
Not bioaccumulated	Some may bioaccumulate
Lung is primary target organ (except CO)	Many target organs
Human health effects readily available	Human dose-response data rarely available
Effects generally occur from minutes to months after exposure	Effects generally occur after long latent period (years)

[a]As regulated under the Clean Air Act, except lead.

in the parts per billion level. Indeed, it has only been with the advent of modern technology that we have been able to regularly and routinely monitor these trace contaminants in the ambient air.

The criteria pollutants are not bioaccumulated in tissues, whereas some of the others do have the tendency to bioaccumulate. This may have significant impacts on long-term health with respect to air toxics.

The lung is the primary target organ for criteria pollutants (with the exception of carbon monoxide). The noncriteria or hazardous air pollutants, on the other hand, have potentially many target organs.

Human health effects data are readily available for the criteria pollutants since they have been studied not only in the ambient air (in some cases for over 100 years), but also in occupational exposures. With respect to the HAPs there is much less human health data available, particularly for dose-response relationships of carcinogens, mutagens, and teratogens.

Finally, the effects of the criteria pollutants tend to last a period of minutes to months, whereas the HAPs tend to have long-term effects.

Acute vs Chronic Effects

It is important to realize that there are two time-related categories of health effects. The first deals with *acute* effects; that is, those health effects which tend to act immediately on a specific target organ or point of entry into the human body. In the air pollution field these are typically the eyes and the lungs, since they are in immediate contact with the ambient air. Burns and asphyxiation are other examples of acute health effects. It is possible for a contaminant to have an acute effect which is different from its chronic effect.

Chronic (Greek, *chronos*) effects are those which refer to time functions of exposure. Thus, there may be a long-term exposure or a long period between an exposure and the resultant health effect. In general, this latter term refers to a more specific concept termed *latency*.

Criteria Air Pollutant Effects

Of the criteria pollutants, ozone, sulfur dioxide, fine particulates, and nitrogen dioxide have both acute and chronic health effects. Carbon monoxide has acute effects only, whereas lead, being a toxic metal, has chronic effects (at ambient air levels).

Ozone

Ozone is a strong oxidizer which affects the respiratory system, leading to damage of lung tissues. Among the acute effects are cough and chest pain, eye irritation, headaches, lung function losses, and asthma attacks. Within the lung itself there is damage to the ciliated cells. These are responsible for the clearance mechanism for particulates in the lungs. In addition, there is damage to the alveoli cell membranes. These are the individual air sacks in the lung where the exchange of oxygen and carbon dioxide take place between the air and the blood.

Chronic exposures to elevated ozone levels are responsible for losses in immune system functions, accelerated aging and increased susceptibility to other infections. In addition, due to its nature as an oxidizer, there are prospects for permanent loss of the alveoli cells.

Sulfur Dioxide

Sulfur dioxide has its own acute health effects, again with the lungs being the target organ. These include irritation and restriction of air passages. There is reduced mucous clearance from the restricted air passages and chest tightness. Otherwise healthy individuals may also experience sore throats, coughing, and breathing difficulties in addition to noticeable odors at concentrations approaching 0.5 ppm. There are some indications of an increased sensitivity to sunlight due to acute exposures.

As for chronic effects, sulfur dioxide is responsible for immune system suppression and increased probabilities of bronchitis. The latter is of particular concern for individuals with emphysema. There are some indications that chronic exposures to sulfur dioxide may also act as a cancer promoter, in addition to being an immune system suppressor.

Particulate Matter

Among the acute health impacts of elevated concentrations of fine particulates are increases in mortality rates, increased incidences of asthma and bronchitis, and increased rates of infection in the respiratory system. They also directly irritate the respiratory tract, constrict airways, and interfere with the mucus lining of the lung passages.

Among the chronic effects of fine particulates are losses of lung capacity and lung damage due to scarring when fine particulates are not cleared from the lung passages or alveoli. In addition, fine particulates act as carriers for toxic contaminants; in particular, heavy metals. This occurs when these materials exist in a fume or a vapor state and condense out in the fine particulate range. In the alveolar regions heavy metals may be absorbed into the blood and circulated to other parts of the body.

Particulates, in particular the fine particulate fractions, are also responsible for visibility reduction. Visibility loss may be considered a psychological stress.

Nitrogen Dioxide

In addition to participating in the formation of photochemical ozone at ground level, nitrogen dioxide has its own particular health effects.

The acute effects of nitrogen dioxide are both direct and indirect. The direct effects are damages to the cell membranes in the lung tissues as well as constriction of the airway passages. Asthmatics are in particular affected by these acute effects. The indirect effects are that nitrogen dioxide causes edema, or a filling of the intercellular spaces with fluid which develops into local areas of infection.

Among the chronic effects of long-term exposures to nitrogen dioxide are necrosis, a term for direct cell death. In addition there are evidences of NO_2 causing a thickening of the alveolar walls of the lungs which interfere with efficient oxygen and carbon dioxide exchange across those cell walls. There appears to be a correlation to increased susceptibility to other lung diseases by chronic exposures.

Some evidence indicates that when mice were exposed to nitrogen dioxide and injected with cancer cells, they developed more cancerous lung nodules than mice injected with cancer cells but who breath clean filtered air. Other studies of NO_2 exposure on mice indicate that changes in lung tissue structure are similar to those which occur in human lungs in the early stages of emphysema. There also appears to be evidence that other target organs may be affected by nitrogen dioxide. The University of Southern California conducted experiments which demonstrated that inhaling NO_2 enlarges the spleen and lymph nodes in mice. The spleen and lymph nodes are important organs in the body's defense system. This finding suggests that the body's immune system may be adversely affected by exposures to chronic levels of nitrogen dioxide.

Carbon Monoxide

Carbon monoxide works through its effect on hemoglobin in the blood. Hemoglobin is the oxygen-carrying protein which is responsible for the oxygen and CO_2 exchanges necessary for life. At high levels of carbon monoxide the potential exists for asphyxiation. In addition, there is impairment of performance, slow reflexes, fatigue, and headaches due to the lack of oxygen in the brain. There are aggravated heart and lung disease symptoms with elevated readings of carbon monoxide and impairments in the central nervous system and brain functions.

The concerns for carbon monoxide are even greater at higher elevations where the partial pressure of oxygen is lower and where many persons may already suffer from an inadequate oxygen supply. Carbon monoxide would further lower the oxygen-carrying capacity of the blood and lead to more rapid acute effects of carbon monoxide.

Lead

Lead, being a toxic metal, acts differently than the other criteria pollutants. It is a systemic toxicant and therefore damages a number of different target organs in the body. Because it is a metal, it is distributed throughout the body and can be responsible for central nervous system damage. Brain functions affected include behavioral changes, losses of muscle control, and learning difficulties. These are the most important aspects of lead exposures.

In addition, lead affects certain key enzymes in the production of red blood cells, which brings on anemia. This is the most characteristic symptom of lead exposure in both children and adults. In addition, there is evidence of kidney damage, liver and heart damage, and damage to the reproductive organs. A combination of effects, including damage to enzyme systems, is possible in these target organs.

BASIC PRINCIPLES OF TOXICOLOGY

A basic understanding of the elements of toxicology is important in order to understand air pollution health effects. The majority of this information deals with animal studies. Similarities and differences, of course, are drawn to epidemiology studies and true human exposures in laboratory settings.

Dose-Response

Toxicity tests, and what we learn from them, are at the heart of under-standing health effects. Since it is illegal to expose people to suspected toxicants, other species are used to quantify the effects that may be expected from exposure to hazardous or toxic substances.

A generalized dose-response curve is seen in Figure 2.1, which could be the response of certain test animals to a chemical. From this curve one may determine the LD_{50} for those animals to that chemical. (LD_{50} is the lethal dose for 50% of the test animals.) Of particular importance is the shape of the curve and the implications thereof. First, it is an S-shaped curve, which indicates that effects are different for incremental changes in dose. Second, it does *not* go through the origin (zero response at zero dose). This indicates that the true effect at very low doses is unknown (or zero for some minimal dose). These important points are examined in detail below.

For the lower dashed line in the figure, one may see essentially no response until some moderate level of dose is reached, at which point the effect becomes truly observable. This is the *threshold* concept and is observed for virtually all materials. The upper dashed line is a linear extrapolation between the lower test points and the origin on the curve. This extrapolation presumes that there will always be a response for a given dose. All other extrapolations are based on assumptions on the part of the investigator.

Major species differences do exist, and this presents the biggest problem in trying to extrapolate dose-response information at all dosages. Absorption rates, metabolic activity, and excretion rates are all included in the

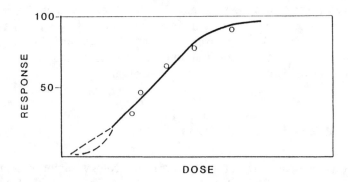

Figure 2.1 Generalized dose-response curve showing low dose extrapolations.

differences between species. Individual differences in the same species can have a dramatic effect on the response of the test animal. Habitat (e.g., aquatic vs terrestrial) is also important. Such features as genetic traits, sex and hormones, nutrition, and age of the test animals in some degree give a scatter in the response data to a given dose. Rats are typically used because they are cheap, have a relatively short life span, and are mammals.

The four most important parameters in comparing dose-response curves and attempting to draw conclusions are species, dose, time period, and end point or response chosen. The biggest problems in interpretation are related to interspecies differences in response to any given chemical.

ROUTES OF EXPOSURE

The four major routes of exposure to hazardous chemicals are inhalation, ingestion, dermal absorption, and injection. Within the environmental field, the latter is a minor route of exposure to environmental contaminants as compared to the other three. Since injection is most often used with experimental animals, the results of experiments using animal tests may not be comparable with the effects from other routes of exposure.

Inhalation

Inhalation, the major route of entry when dealing with hazardous air pollutants, involves the intake of airborne chemicals during breathing. The solubility of the material in the blood affects the degree of its absorption. Once in the blood via the lungs, it goes directly to the brain and the rest of the body.

The most common example of environmental exposure by this route is the absorption of carbon monoxide through smoking. Carbon monoxide has several hundred times the binding affinity for hemoglobin as does oxygen. Hemoglobin is the iron-based organic compound that is responsible for all oxygen transport in the blood. Oxygen is "bound" by hemoglobin and is later given up by hemoglobin to the tissues. When oxygen is displaced irreversibly by carbon monoxide, the transport phenomenon is not able to function and oxygen starvation of cells begins.

The retention of airborne particulates that are, or may carry, toxic chemicals is highly dependent upon particle size due to the structure of the human lung. The smaller sizes penetrate deeper and have a greater effect.

Table 2.3. Systemic Poisons

Hepatotoxic agents (liver)	Neurotoxic agents (nerve system)
Carbon Tetrachloride	Methanol
Tetrachloroethane	Carbon Disulfide
	Metals (Pb, Hg)
Nephrotoxic agents (kidneys)	Organometallics
Halogenated Hydrocarbons	Benzene
Hematopoietic toxins (blood)	**Anesthetics/narcotics (consciousness)**
Aniline	Acetylene Hydrocarbons
Toluidine	Olefins
Nitrobenzene	Ethyl Ether, Isopropyl Ether Paraffin
Benzene	Hydrocarbons
Phenols	Aliphatic Ketones
	Aliphatic Alcohols
	Esters

Source: Clayton, G. D., and F. E. Clayton, *Patty's Industrial Hygiene and Toxicology,* Vols. 1 and 2 (New York: John Wiley and Sons, Inc., 1986).

RESPONSE TO AIRBORNE CHEMICALS

Chemical agents have a different effect on different member organs of the body, depending on the dose and route of exposure. Other chemicals target specific organs, depending on oil or fat solubility, the effect that the chemical may have on enzyme activity, or physical interruption of the transmission of electrical impulses.

Table 2.3 lists a variety of systemic poisons that influence different target organs in the human body by type of chemical or hazardous substance. The hepatotoxic agents, such as carbon tetrachloride, primarily affect the liver, whereas nephrotic agents (halogenated hydrocarbons) affect the kidneys. The neurotoxic agents that affect the nerve system include methyl alcohol, carbon monoxide, heavy metals, and organometallic compounds. The hematopoietic toxins affect the blood or blood cells and consist of aromatic compounds such as benzene, phenols, aniline, and toluidine. The anesthetic or narcotic chemicals (which affect consciousness) consist of ketones, aliphatic alcohols, and double-bonded or "ether" types of organic compounds.

The Lungs

The lungs are the most quickly impacted of all of the body organs to toxic contaminants. This is due to the fact that the lungs are in constant contact with the environment. Also, the lungs have a surface area of 70 – 100 m², as opposed to the skin at 2 m², or the intestinal tract at about 10 m². This becomes an important point since the rates of absorption of various contaminants are considered to be a direct function of the surface area exposed to a contaminant. Among the criteria pollutants which have a direct and adverse impact on the lungs are ozone, sulfur dioxide, nitrogen dioxide and fine particulates.

Figure 2.2 presents the various passages and sections of the human lung. The lung acts as a particle filter by its varied construction. As one goes

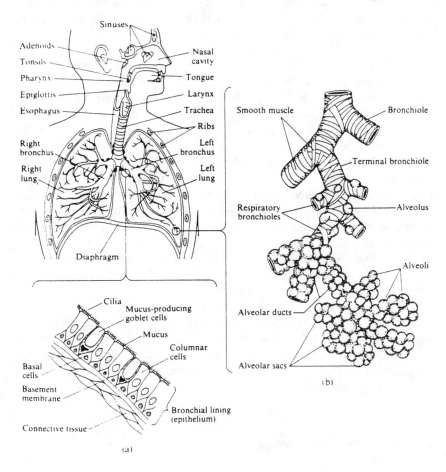

Figure 2.2   (a) Air passages of the respiratory system of man; (b) Details of the terminal respiratory units of the lung.

deeper in the lungs, one finds more narrow and torturous paths which present an aerodynamic obstruction to particle movement. Particle size, therefore, determines where particles will be deposited in various portions of the lung (the smaller, the deeper). Gases penetrate to the deepest segments of the lung.

The nose and pharynx will capture particles from 5 μm to greater than 50 μm in diameter. The larynx down to the bronchi regions of the lungs collect particles from 1 to 5 μm in size. The smaller portions of the lungs, the bronchioles and alveoli, will collect particles in the 0.5 μm range. (Particles smaller than 0.05 μm will be exhaled.) The filtering construction of the lung is helped by the fact that mucus covers the upper reaches of the lung passages. Particles that are deposited in the branched passages of the upper reaches of the lung will be removed by a phenomenon called the "mucociliary elevator." This is a movement of mucus upward and out of the lungs caused by small ciliary hairs in the lung passages. Thus, inhaled material will be moved up out of the lungs by this "elevator" effect after the material has been captured in the mucus. (Note that inhaled material could potentially reach the stomach from the lungs due to a swallowing action.)

Down in the alveoli portions of the lungs, only the small particles can directly enter. But once in, they do not depart—the alveoli are not coated with mucus. Damage may thus occur in the most important part of the lung. Inflammation and irritation are also possible, which may cause the lungs to be scarred and toughened. The alveoli or tiny air sacs may lose their elasticity due to repeated impact by particles and the scarring action of gases, paints, or solvents. Emphysema is one result. There are approximately 200 million alveoli in the average lung; however, they do not regenerate once they have been formed. Therefore, if they are destroyed as a result of air pollution, those alveoli are never regained. Thus lung capacity is lost and the lung's ability to function is decreased.

Certain contaminants or even small dissolvable particulates may enter the blood stream directly and then be transported to target organs such as the liver or kidneys for their effect. Lung cancer is the one effect which has been established as being produced by asbestos particles. This case has been documented so that some models of cancer-causing mechanisms have a basis in reality.

The Central Nervous System

The central nervous system (the brain, spinal column, and nerves) is generally impacted by heavy metals such as mercury or lead. Air pollutants, such as carbon monoxide and aromatic compounds (benzene) directly affect the central nervous system by asphyxiation due to brain oxygen and blood loss.

Minimata disease is a prime example of central nervous system damage caused by exposure to a contaminant, as is the "Mad Hatter's disease" of the 19th century. The Mad Hatter in *Alice in Wonderland* suffered, as did most hatters, from mercury poisoning caused by a high level of mercury in the tanning solutions used to make hats for men and women.

The Liver

The liver is a serious target organ, because it is a metabolic center of the entire body. For example, it is in the liver that carbon tetrachloride is converted to chloroform, which is a carcinogen and toxic to the cells with which it comes in contact. Sufferers from other liver dysfunction diseases, such as alcoholism, are acutely impacted.

The Kidneys

The kidneys are the main filtering media in the body and thus have a high exposure to toxicants. As a filter, they serve to concentrate certain contaminants, such as heavy metals and halogenated hydrocarbons. As the concentration increases, the "dose" becomes greater and toxic or carcinogenic responses increase.

The Blood

The blood system is impacted by agents such as carbon monoxide that affect the oxygen-carrying capacity of hemoglobin. Direct blood cell impacts also occur due to chemicals such as the aromatic compounds benzene and toluene, and phenolic compounds.

The Reproductive System

The reproductive system is prone to impact by environmental contaminants because it is a center of DNA activity. Thus, any contaminants that affect cell division and the transmission of genes will affect the reproductive system. There are significant differences between the sexes regarding their response to reproductive toxins. Fertility, sperm count, and cancers are all affected by environmental contaminants. Lead and diethyl stilbestrol (DES) have decided effects on the reproductive system.

The Cardiovascular System

The cardiovascular system and, in particular, the heart is directly affected by carbon monoxide by limiting the supply of oxygen in the blood going to the heart muscle. Death of heart muscle tissue is a result which lessens the ability of the heart to perform and acts as a potential precursor to heart failure.

The Skeletal System

Heavy metals such as lead and strontium tend to accumulate in the skeletal system by displacing calcium, whereby they become a reservoir for these contaminants which then may act as systemic poisons. In addition, the potential impacts of radioactive isotopes of both of these metals may lead to direct damage to the skeletal system.

Other Factors to Consider

Table 2.4 gives a relative index of toxicity that may be used when considering the oral lethal dose for human beings. While this is not a specific reference table, one may get some idea of the amount of a toxicant it would take to be lethal to an average human being. Elderly, young, and smaller-bodied individuals would show an enhanced effect over the average.

Table 2.4. Relative Index of Toxicity

Toxicity Rating or Class	Probable Oral Lethal Dose for Humans	
	Dose	For Average Adult
1. Practically nontoxic	> 15 g/kg	More than 1 quart
2. Slightly toxic	5–15 g/kg	Between 1 pint and 1 quart
3. Moderately toxic	0.5–5 g/kg	Between 1 ounce and 1 pint
4. Very toxic	50–500 mg/kg	Between 1 teaspoonful and 1 ounce
5. Extremely toxic	5–50 mg/kg	Between 7 drops and 1 teaspoonful
6. Super toxic	< 5 mg/kg	A taste (less than 7 drops)

Individual differences may thus have a profound effect on response to toxic chemicals. Genetic makeup controls the presence or absence of key enzymes that are the biochemical catalysts for metabolism. Humans are highly heterogeneous, and individual differences are widespread.

Sex and hormones make for different responses to toxicants. Hormones influence enzyme levels, and therefore chemical toxicity. Thus, males and females are often not equally sensitive to chemicals, and pregnant females are in some instances affected differently than nonpregnant females. Differences in physiology between males and females can also influence response. For example, females have a higher average body fat content, and may be more adversely affected than males by fat-soluble contaminants such as PCBs or DDT.

Infants and children are often more sensitive due to undeveloped tissues, and possess a reduced ability to metabolize and detoxify chemicals. Additionally, they may have a much lower body mass, so for a given intake the dose would be much higher than for an adult. The elderly may be more sensitive to toxic chemicals due to the reduced detoxifying capacity of the liver and excretory capacity of the kidneys. Susceptibility to injury or other aging factors may also increase sensitivity to environmental toxicants.

The different routes of exposure to a chemical carcinogen or toxicant will have a decided effect on how the body reacts. For instance, nickel fumes are carcinogenic if inhaled through the nose or larynx, but if nickel fumes are ingested, they are not carcinogenic.

CLASSES OF HEALTH EFFECTS

There are a wide variety of bodily responses to environmental agents that are of concern. The following are six classes of toxic agents, examples of each and the bodily responses they provoke:

- allergic agents—isocyanates: cause itching, sneezing, or rashes
- asphyxiants—such as cyanide and carbon monoxide: displace oxygen
- irritants—such as hydrochloric acid, ammonia, and chlorine: may cause pulmonary edema at very high concentrations, or unpleasant sensations throughout the body
- necrotic agents—ozone and nitrogen dioxide: directly cause cell death
- carcinogens—asbestos, arsenic, and cigarette smoke: cause cancer
- systemic poisons—benzene and arsenic compounds: attack the entire body

Table 2.5. Synergism of Asbestos and Smoking

Group	Asbestos Exposure	Cigarette Smoking	Death Rate[a]
Control	0	0	11
Asbestos workers	Yes	0	58
Control	0	Yes	123
Asbestos workers	Yes	Yes	602

[a]Lung cancer death rate per 100,000 person years.

Synergism and antagonism are two concepts important to the understanding of toxic effects in the human body. Synergism is best summarized by saying two plus two equals five. This describes the enhancing effect of two different environmental contaminants upon the human body. For instance, asbestos may cause cancer, but the carcinogenic effect of asbestos is many times greater with a person who also smokes.

Table 2.5 summarizes this effect from epidemiology studies on workers with asbestos and compares their lung cancer death rates with and without cigarette smoking. For the control group, the death rate was 11 per 100,000 person years, whereas with asbestos only the death rate went up to 58. Smoking alone had a death rate of 123. When both asbestos and cigarette smoking were present in this epidemiology study it was found that the death rate jumped by a factor of approximately 10 times when cigarette smoking was present in addition to asbestos.

Antagonism also occurs, and is best summarized by saying two plus two equals three. Such an antagonistic effect is where one impact is lessened by having another substance present. Verifiable examples of antagonistic effects are difficult to find in the research literature on epidemiologic studies.

Latency

A concept important to toxic effect is the latency period. This is the time between exposure of an individual and the clinical manifestation of an adverse effect. Mesothelioma (a chest cancer) is a prime example of latency. Workers involved in shipbuilding during World War II were exposed to airborne asbestos fibers at very high concentrations, yet the effect (cancer) did not show up until 20 to 30 years after the exposure had ceased.

Specific Contaminant Effects

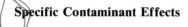

Carcinogens

Carcinogens are those contaminants that may cause cancer. On the average, cancers are approximately 15% related to genetic effects and 85% related to environmental effects (including both lifestyle and environment).

Mutagens

Mutagens are agents that cause transmittable changes in genetic material within an organism. In humans, these agents alter the gene structure of the DNA in somatic and/or germ cells which can affect subsequent generations. The most powerful mutagenic effects are known to come from radiation overexposures.

Teratogens

Teratogens are agents that adversely affect offspring while in the fetal stage of development. Thus, pregnant women would be most likely to experience a teratogenic effect due to exposure to a hazardous air pollutant. A well-documented case of such a teratogenic effect was the birth defects found in children of women who ingested thalidomide-containing sleeping pills in the early 1960s. This effect was noted only when the pregnant women took thalidomide on the 12th day following fertilization. Thus, it has a developmental effect on offspring during the fetal period. Other teratogenic substances include alcohol, carbon monoxide, anesthetic gases, and DES.

EFFECTS ON VEGETATION AND CROPS

Widespread injury to agricultural crops was reported as early as the mid-1940s in Los Angeles. These losses were due to phytotoxic air pollutants such as O_3 and peroxyacyl nitrate (PAN). Spinach was the first crop to be lost to air pollutant damage in Los Angeles.

Since these early discoveries, O_3 injury on sensitive vegetation has been observed in many parts of the United States. Levels sufficient to cause injury to very sensitive vegetation are reported in most areas east of the Mississippi River.

Pollutants which have a long history of being known to cause significant plant injury under ambient conditions of exposure include SO_2, fluorides, O_3, PAN, and ethylene. Although it has been suspected to be a major plant-injuring pollution problem for some time, acidic deposition has only recently been correlated by some scientists to plant injury under ambient conditions (e.g., the decline of red spruce in high altitude forests). Other pollutants are also known to cause injury to plants. These include NO_2, HCl, and particulate matter.

Plant Structure

Plant injury is dependent on a number of physical and biological factors. Plant structure and anatomy have a significant influence on plant response to air contamination. A plant consists of four organs: roots, stems, leaves, and reproductive structures. The leaf is the principal target, since it is the organ involved in gas exchange and its damage is most obvious. Although leaf function is consistent from species to species, significant differences in leaf structure and anatomy exist.

Leaf Structure

The upper surface of a leaf is overlain by a waxy layer referred to as the cutin. Below the cutin is a layer of colorless cells, the upper epidermis. Both the cutin and the upper epidermis protect the leaf from desiccation and mechanical injury. Located beneath the upper epidermis is a layer of photo-synthetically active cells. Next is a mass of irregularly shaped and loosely arranged cells. The loose arrangement provides for large intercellular spaces where gas exchange is facilitated. The lower surface of the leaf is bounded by the lower epidermis, which also functions to protect the leaf.

In most species, large numbers of openings called stomata are located in the lower epidermis. In some species stomata are located on both the lower and upper surfaces. The stomata consist of a variable-sized pore flanked by two crescent-shaped guard cells. The constriction or relaxation of the guard cells determines the pore diameter and regulates the rate of gas exchange for CO_2 and H_2O. It is through these stomata that pollutant gases enter and react with the tissues of the leaf. The ambient temperature, humidity, and carbon dioxide levels determine the degree to which the stomata constrict.

Plant Injury

Visible effects are identifiable changes in leaf structure, which may include chlorophyll destruction (chlorosis), tissue death (necrosis), and pigment formation. Visible symptom patterns may result from acute or chronic exposures. Acute injury often results from brief exposures (several hours) to elevated levels of a pollutant. Tissue necrosis is generally the dominant symptom pattern from acute exposures.

Chronic injury usually results from intermittent or long-term exposures to relatively low pollutant concentrations, with chlorophyll destruction or chlorosis as the principal symptom of injury.

Subtle effects imply that no visible injury is apparent. Such effects can only be discerned by measurements of physiological processes such as photosynthesis or overall growth reduction as measured by dry cell mass.

The severity of the injury is dependent on the dose to which the plant has been exposed. The greater the dose, the more severely injured are individual leaves and the whole plant. The entire leaf, or in some extreme cases the entire plant, may be killed.

Acid Precipitation Impacts

Acidic materials deposited on plants and the soil have the potential for causing injury or measurable changes in plants. Laboratory and greenhouse studies with *simulated* acidic rain events have shown that a wide variety of plants can be injured by exposure to solutions with a pH of approximately 3.0. Field observations of specific damages due to acid rain are not verified. Lab symptoms reported include pitting, necrotic lesions, chlorosis, wrinkled leaves, and marginal and tip necrosis. The most common symptom of acid injury on plants observed in a wide variety of studies is small (< 1mm) necrotic lesions or "burn spots."

Other effects (not manifested as visible injury) include changes in yield. Simulated acid rain effects on yield have resulted in a number of outcomes. For example, the yield of crops such as tomatoes, green peppers, strawberries, alfalfa, orchard grass, and timothy were observed to be *stimulated!* Yields of broccoli, mustard greens, carrots, and radishes were inhibited. In other crop plants, yields were unaffected.

Acidic deposition may also affect plants indirectly by its effects on the chemistry of the soil. These effects may be either positive or negative. Increased growth or yield of plants in some studies has been attributed to the fertilizing effect of nitrates and/or sulfates. The addition of nitrates to nitrogen-deficient soils has been suggested to stimulate forest growth. Ammonium nitrate and ammonia sulfate are, of course, commercial fertilizers.

Acidic deposition has been implicated in the widespread decline of Norway spruce and other conifers in the Black Forest of Germany. The implication of acidic deposition of some type is due, in part, to the fact that decline of conifers is associated with increasing altitude. Mountain regions are more likely to experience acidic fogs or cloud droplets for over one-third of the year. Leaching of the nutrients from the tree foliage and crown have been implicated in lab studies using acid mists. The true cause is unknown. It may be a combination of many factors.

Pollutant Interactions

Air pollutant synergism also occurs in plant impacts. Therefore, it is possible that simultaneous, sequential, or intermittent exposures to pollutants may result in plant injury. Available evidence indicates that simultaneous exposures to gaseous mixtures can produce synergistic, additive, or antagonistic effects.

Studies of plant exposures to *mixtures* of ozone and sulfur oxides, as well as ozone and NO_2 have reported decreased injury thresholds. Antagonistic responses are generally observed when injury caused by pollutants applied singly is severe; the effect of mixtures is to reduce the severity of injury.

EFFECTS ON MATERIALS

Gaseous and particulate air pollutants are known to significantly affect materials. In the United States these effects cause economic losses estimated to be in the billions of dollars each year. Of particular importance are effects on metals, building stones, paints, textiles, fabric dyes, rubber, leather, and paper. Significant effects on these materials have been observed in other industrialized nations as well.

Materials can be affected by both physical and chemical mechanisms. Physical damage may result from the abrasive effect of dust deposition. Chemical reactions may result when pollutants and materials come into direct contact. Absorbed gases may act directly on the material, or they may first be converted to new substances that are responsible for observed effects. The action of chemicals on materials usually results in irreversible changes.

Textiles

Exposures to atmospheric pollutants may result in significant deterioration and weakening of textile fibers. Fabrics such as cotton, hemp, linen, and rayon, which are composed of cellulose, are particularly sensitive to acid aerosols and acid-forming gases. Synthetics such as nylon may also incur significant acid damage. The apparent disintegration of women's nylon hose in urban environments heavily polluted by SO_2 and acid aerosols received considerable notoriety in the past. Nylon polymers may also undergo oxidation by NO_2, which reduces the affinity of nylon fibers for certain dyes.

Air pollutants may also react with fabric dyes, causing them to fade. Such fading has been associated with SO_2, particulate matter, NO_2 and O_3. Fading of textile dyes associated with nitrogen oxide exposure, particularly NO_2, has a long history.

Reports of NO_2-induced fading of textile materials from ambient NO_2 exposures were soon followed by observations that ambient O_3 levels were also a prime cause of fading. In the early 1960s, O_3 fading was reported for polyester-cotton permanent press fabrics and nylon carpets.

Some manufacturers have suffered significant economic losses. Fading of nylon carpets was a problem along the Gulf Coast. Blue dyes were particularly sensitive to a combination of ambient O_3 exposure and high humidity. This "fading" was overcome by using O_3-resistant dyes and modification of the nylon fibers to decrease the accessibility of O_3.

Building Materials

In addition to soiling, building materials such as marble, limestone, and dolomite may be chemically eroded by acidic gases. This erosion is caused by exposures to acidic gases in the presence of moisture, acid aerosols, or acid precipitation. The reaction of sulfuric acid with carbonate building stones results in the formation of $CaSO_3 \cdot 2H_2O$ and $CaSO_4 \cdot 2H_2O$ (gypsum), both of which are soluble in water.

These effects are not limited to the surface, as water transports acids into the interior of the stone. The soluble salts produced may precipitate from solution and form incrustations or they may be washed away by rain. The chemical erosion of priceless, irreplaceable historical monuments and works of arts in Western Europe are examples of these damages.

The Colosseum and the Taj Mahal are in various stages of dissolution. In many cities of Europe marble statues have had to be moved indoors.

Recently, the bronze statue of Marcus Aurelius, which stood for nearly two millennia in Rome, had to be removed permanently from the Capitoline Hill due to air pollution.

A significant concern is the Acropolis, located downwind of the city of Athens, Greece. Although the chemical erosion of the marble structures in the Acropolis has been occurring for over a hundred years, the rapid population growth and use of high-sulfur heating oils since World War II has greatly accelerated their destruction.

Metal Corrosion

The corrosion of metal in industrialized areas represents one of the most ubiquitous effects of atmospheric pollutants. When iron-based metals corrode, they take on the characteristic rust appearance. From a variety of studies it is apparent that the acceleration of corrosion in industrial environments is associated with SO_2 and particulate matter. The probable agent of corrosion in both instances is sulfuric acid produced from the oxidation of SO_2.

Nonferrous metals may also experience significant pollution-induced corrosion. For example, zinc, which is widely used to protect steel from atmospheric corrosion, will itself corrode when acidic gases destroy the basic carbonate coating that normally forms on it. The reaction of SO_2 with copper results in the familiar green coating of copper sulfate that forms on the surface of copper coated materials.

Because nonferrous metals are used to form electrical connections in electronic equipment, corrosion of such connections by atmospheric pollutants can result in serious operational and maintenance problems for equipment users.

Pollutants may also damage low power electrical contacts used in computers, communications and other electronic equipment by forming thin insulating films over contacts. Such films may result in open circuits, causing the equipment to malfunction. Equipment malfunction may also result from the contamination of electrical contacts by particulate matter. Particles may physically prevent contact closing or they may result in chemical corrosion of contact metals.

Surface Coatings

The function of surface coatings is to provide a protective film over solid materials to protect the underlying material from deterioration. Paint appearance and durability are affected by air pollutants such as particulate matter, H_2S, SO_2, and O_3. Particles may also serve as wicks that allow chemically reactive substances to reach the underlying material, resulting in corrosion if the underlying material is metallic. Pollutant effects on paints may include soiling, discoloration, loss of gloss, decreased scratch resistance, decreased adhesion and strength.

House paints pigmented with lead may be discolored by reaction of the pigment lead with low atmospheric levels of H_2S. The intensity of the discoloration is related not only to the concentration and duration of the exposure but also to the lead content of the paint and the presence of moisture.

Documents

Paper is sensitive to SO_2. This sensitivity is due to the conversion of SO_2 to sulfuric acid by impurities in the paper. Sulfuric acid causes the paper to become brittle, decreasing its service life. This embrittlement is of concern to libraries and museums in environments with high SO_2 levels, as it makes the preservation of historical books and documents much more difficult. Most historical documents today, the Magna Charta, the Declaration of Independence, etc., are therefore kept totally isolated from the atmosphere.

Pollution damage to leather is similar to that of paper, as both are apparently caused by sulfuric acid produced from SO_2.

Rubber

Ozone can induce cracking in rubber compounds that are stretched or under pressure. The depth and nature of this cracking depends on the O_3 concentration, the rubber formulation, and the degree of rubber stress. Unsaturated natural and synthetic rubbers such as butadiene-styrene and butadiene-acrylonitrile are especially vulnerable to O_3 cracking. Ozone attacks the double bonds of such compounds, breaking them when the rubber is under pressure. Saturated compounds such as thiokol, butyl, and silicon polymers are resistant to O_3 cracking. Unsaturated compounds which are chlorinated, such as neoprene, are also resistant. Such resistance is necessary for rubber in products as automobile tires and electrical wire insulation.

EFFECTS ON ANIMALS

Domesticated animals such as dogs and cats are also affected by air pollution. Fluoride has caused more confirmed air pollution injury to domesticated animals than any other air pollutant. Most cases of fluoride toxicity (fluorosis) have resulted from the contamination of forage. It has been forage-consuming livestock animals such as cattle, sheep, horses, and pigs that have been the most commonly poisoned by fluoride.

Fluorosis can be either acute or chronic. Acute fluorosis is rare since livestock will not voluntarily consume heavily contaminated forage. Chronic fluorosis is, however, commonly observed in livestock which ingest fluoride-contaminated forage over a period of time. As fluoride interferes with the normal metabolism of calcium, chronic fluoride toxicity is characterized by dental and skeletal changes. One of the earliest signs of fluorosis is white chalky patches or mottling of dental enamel. Skeletal changes, such as calcification of ligaments and thickening of the long bones, also occur, as does decreased milk production.

ECONOMIC LOSSES

Economic losses due to anthropogenic air pollution-induced damage to materials are difficult to quantify since one cannot easily distinguish what is due to the natural deterioration of materials and what is caused by mankind.

Estimates of material damages in the U.S. run up to $2 billion annually for textiles and fabrics, up to $1 1/2 billion for metals, and $500 million each for paints and rubber products.

Economic Losses Due to Vegetation Impacts

Excluding the photochemical oxidant problem in Southern California and some areas of the Northeast, most reports of air pollution-induced plant injury have been associated with point sources, with injury being localized. With the exception of primary metal smelters, economic losses associated with point sources have often been insignificant. Nevertheless, because of photochemical oxidants such as O_3 and PAN, air pollution injury to vegetation is still widespread. These photochemical oxidants continue to cause significant injury to plants and economic losses in many regions of the country. A number of attempts have been made to estimate the annual economic losses to crops.

These loss estimates did not take into account the widespread damage to ponderosa pine and to sensitive trees. Based on a California survey, a $135 million annual extrapolated loss to crops has been projected for the United States.

Scientists who conduct research on the toxic effects of air pollutants on plants generally agree that O_3 causes 90% or more of the air pollution damage to crops in the United States.

3   AIR QUALITY STANDARDS AND MONITORING

It has been believed for some time that ozone or some material resembling it is present. . . . Rubber bands placed on the roof of the laboratory in Los Angeles exposed to the atmosphere show rapid deterioration. . . . .

The Smog Problem in Los Angeles,
the Committee on Smoke and Fumes — WOGA, 1949

Acceptable levels of air quality are important in defining health impacts or potential risks to the population at large. Likewise, how well we meet ambient air quality standards becomes the measuring tool by which we judge air quality management strategies. Figure 3.1 indicates the severity of air pollution experienced by different cities around the world. Shown are the second highest one-hour ozone and the annual particulate and sulfur dioxide concentrations experienced in a number of locations. These do not tell the entire story, but they do indicate that certain regions of the world are subject to severe concentrations of ozone, particulate matter and sulfur dioxide.

Air quality standards are expressed in terms of a given concentration of the contaminant over a specified period of time. The dose concept for any contaminant is therefore an integral part of evaluating and/or expressing levels of acceptable air quality. When the air quality standards are set, they generally incorporate the dose concept as the basis for the most relevant health effects. The type of health effect and exposure time are also important. Ozone usually has a short duration standard. Lead, on the other hand, is expressed almost exclusively by monthly or annual average concentrations

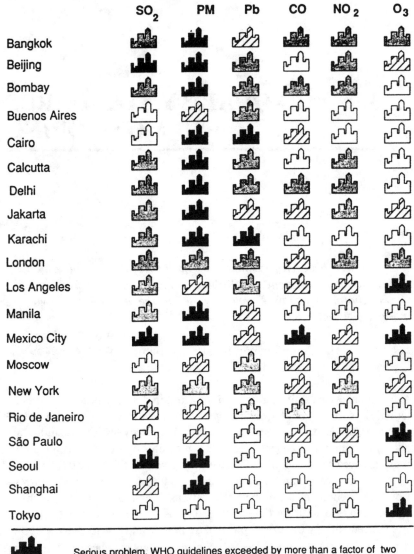

Figure 3.1.   Relative air quality worldwide.

throughout the world. This is due to the chronic health effects of lead being the most prominent.

TYPES OF AIR QUALITY STANDARDS

The criteria air pollutants in the United States are ozone, carbon monoxide, nitrogen dioxide, sulfur dioxide, particulate matter, and lead. These contaminants not only have a primary standard which is designed to protect human health but, in some cases, a secondary standard which is to protect the environment (materials, vegetation, etc.). Secondary ambient air quality standards for the United States federal government are the same as the primary standards, with the exception of sulfur dioxide. These are the pollutants designated by the federal government for which national ambient air quality standards (NAAQS) have been set and are in effect for all states, territories, and United States possessions. Individual states may adopt *more stringent* air quality standards of their own.

The noncriteria air pollutants in general have two bases upon which standards may be set. These include occupational exposure level (OEL) based standards and risk based standards. The OEL based standards for HAPs utilize some published value for a known health effect, such as the threshold limit values as the starting point.

For these OEL based levels, regulatory agencies typically apply a factor to established acceptable levels consisting of fractions between 1/10 and 1/1000 as "corrections" to allow increased sensitivities for the general population and susceptible population groups. Thus, an "acceptable" level of air quality for noncriteria air contaminants, using the OEL based system, could be derived from a TLV divided by a factor of ten to a thousand, which the regulatory body would consider acceptable.

Another approach to acceptable air quality for toxic air contaminants are those that are based upon levels of individual and/or societal risk. These latter approaches form the basis of the health risk assessment methodology considered later. These are risk-based approaches specifically for those airborne contaminants which are carcinogenic, mutagenic, and/or teratogenic.

Contaminants which cause a *public nuisance,* primarily due to odors, do not have a numerically-based threshold or acceptable concentration, but are derived from public complaints. This lack of numerical values is because odor thresholds vary significantly among the population at large. Public nuisance, however, is among the oldest of all air quality standards and, indeed, is the only situation for which there is no numerical value. Therefore, standards based upon odors or public nuisance commonly require a number of individuals to petition a regulatory body for relief.

Acceptable Levels

Concerns for acceptable air quality range from life-threatening levels to occupational exposures, general population standards, and exposures corresponding to a level of cancer risk. Figure 3.2 illustrates relative acceptable concentrations for each concern.

Both the National Institute for Occupational Safety and Health (NIOSH) and the Occupational Safety and Health Administration (OSHA) have derived for workers an immediately dangerous to life and health (IDLH)

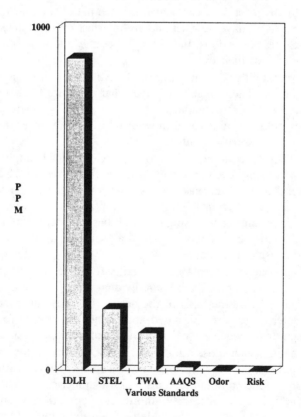

Figure 3.2.   Relative inhalation levels.

level for emergency situations for a variety of contaminants. This is the maximum concentration from which one could escape within 30 minutes without experiencing an escape-impairing or irreversible health effect. Carcinogens do not have an IDLH.

The best known of the allowable exposure levels in the US for occupational standards are the threshold limit values (TLVs) published by the American Conference of Governmental Industrial Hygienists. These are levels of airborne contaminants below which healthy adults, working for 40 hours per week, generally do not suffer adverse health effects due to repeated exposures. Unless adopted by a governmental agency, these are not government standards but are recommendations made by industrial hygienists. These limit values are updated annually.

The threshold limit values are published for two time intervals. The *time-weighted average* (TWA), is an allowable eight-hour average exposure concentration, and the short-term exposure level (STEL), the maximum 15-minute average concentration to which workers may be exposed. OSHA has adopted certain of these TLVs as permissible exposure levels (PELs). PELs, therefore, become a legally enforceable level of exposure for industrial workers.

Other toxicological levels which may be used include the "No Observed Adverse Effect Level" (NOAEL). The NOAEL is frequently used as a starting point of allowable exposure for potential human inhalation levels. Other acceptable level approaches use the "No Observed Effect Level" (NOEL), which includes data published in the literature as an equivalent health level. The latter two levels are based upon animal studies, whereas the TLVs are based upon human studies.

As an illustration, Table 3.1 compares four common air contaminants and their different levels of acceptable air concentration for different situations. Ozone and sulfur dioxide are compared against the hazardous air pollutants vinyl chloride and formaldehyde.

The life-threatening level (IDLH) is 10 ppm for ozone and 100 ppm for sulfur dioxide. Neither vinyl chloride nor formaldehyde have an IDLH level. On the other hand, an eight-hour TLV does exist for all four contaminants. These levels are: 0.1 ppm for ozone, 2 ppm for sulfur dioxide, 5 ppm for vinyl chloride and 0.3 ppm for formaldehyde.

The federal Ambient Air Quality Standard (AAQS) for ozone (0.12 ppm) and sulfur dioxide (0.14 ppm) have different averaging times of 1 hour and 24 hours, respectively. For comparison, the air quality standards adopted by California are also shown. For vinyl chloride, a 24-hour standard of 0.01 ppm has been adopted by that state.

Other levels that may present a concern for public nuisance would be the odor thresholds. With respect to the potential for a cancer risk of one

Table 3.1. A Comparison of Significant Air Pollutant Levels

Level	Ozone	SO$_2$	Formaldehyde	Vinyl Chloride
IDLH	10	100	[a]	[a]
TLV	0.10	2	0.3	5
Federal AAQS[b]	0.12	–	–	–
24 hr.	–	0.14	–	–
Calif. AAQS[b]	0.10	0.25	–	–
24 hr.	–	–	–	0.01
Odor Threshold	0.015	0.5	< 1	–
10^{-6} Cancer Risk	NA	NA	0.14 ppb	0.01 ppb

Concentrations in ppm, unless otherwise noted.

[a] = Carcinogen.

IDLH = Immediately Dangerous to Life or Health (30 min.).

TLV = Threshold Limit Value, 8 hr Time-Weighted Average. 1991/92 ACGIH.

NA = Not Applicable.

[b] = 1 hour Concentration.

chance in a million, there are no standards for ozone or sulfur dioxide. For vinyl chloride and formaldehyde, they exist at exceedingly low concentrations.

AMBIENT AIR QUALITY STANDARDS AND EXPOSURES

In excess of 86 million people were exposed to criteria air pollutants at levels in excess of one or more of the national ambient air quality standards (NAAQS) in 1991. Of those, the majority live in areas in which the ozone standard was exceeded. This indicates that there are valid concerns for health, and that significant numbers of people are impacted by air pollutants.

The national ambient air quality standards establish a concentration criteria for each known health impact threshold. Federal primary standards are those levels of air quality necessary to protect the public health while allowing for an adequate margin of safety. Secondary standards are those levels necessary to protect the public welfare from known or anticipated adverse effects, including economic values and personal comfort.

Table 3.2 lists the federal primary ambient air quality standards. Each of these contaminant levels indicates acceptable air quality. These AAQSs include provisions for sensitive population groups such as the elderly, asth-

Table 3.2. National Ambient Air Quality Standards
Criteria Pollutants

Contaminant	Concentration[a]	Duration (hours)
Ozone	0.12	1
Carbon Monoxide	35	1
	9	8
Nitrogen Dioxide	0.05	annual
Sulfur Dioxide	0.14	24
	0.03	annual
Particulate Matter-PM10	150 $\mu g/m^3$	24
	50 $\mu g/m^3$	annual
Lead	1.5 $\mu g/m^3$	3 months

[a]in parts per million (vol), unless otherwise indicated

matics, young children, etc. At the federal level, air contaminants with a chronic adverse health impact have annual-based standards, whereas air pollutants with more acute effects tend to have one-hour standards. The only secondary AAQS different from a primary standard is the three-hour SO_2 concentration of 0.5 ppm (no three-hour primary SO_2 standard exists). The particulate matter standard (PM10) is based on the mass of particles with an aerodynamic diameter less than 10 microns in diameter.

A comparison of other ambient air quality standards is shown in Table 3.3. This is a representative list of selected national ambient air quality standards for the criteria pollutants, except for lead. Included are the recommended values by the World Health Organization. As we can see, different countries of the world have differing standards as well as different average times for the representative criteria pollutants. For example, Hungary has the lowest ozone standard at 0.03 ppm for a one-half hour average, whereas the highest standard is 0.15 ppm for one hour in Saudi Arabia. Other countries do not have ambient air quality standards, and in some instances the World Health Organization has no recommended standard.

Depending upon the severity of the air pollution problem in a given area, those geographical areas of the United States which exhibit air quality in excess of federal standards are considered *nonattainment* areas. Federal and state regulations mandate certain actions to be taken in order to attain the primary ambient air quality standards first, and then the secondary ambient air quality standards.

The United States is divided geographically into metropolitan statistical areas (MSAs) which are used for evaluating exposures to air contaminants and the population where the AAQSs are not attained. Table 3.4 indicates

Table 3.3. Representative International Ambient AQ Standards (ppm)[a]

Country	O_3[b]	CO[d]	SO_2[e]	NO_2[f]	PM[e,g]
World Health Organization	0.075	9	0.048	–	120
United States	0.12	9	0.14	0.05	150
Switzerland	0.06	7.2[e]	0.038	0.015	150
South Africa	0.12	–	0.10	0.135	150
Saudi Arabia	0.15	9	0.14	0.05	340
Mexico	0.11	13	0.13	–	275
Japan	0.06	20.5	0.04	–	100
Israel	0.115[c]	10	0.11	–	200
Hungary	0.03[c]	4.5[e]	0.17	0.035	100
Canada	0.05	5	0.17	0.03	120

[a]All data rounded off
[b]1 hour average
[c]0.5 hour average
[d]8 hour average
[e]24 hour average
[f]annual average
[g]$\mu g/m^3$, not parts per million

the number of nonattainment areas for the primary ambient air quality standards. Under federal law, the more severe an area is in nonattainment, the more stringent the air quality management requirements become for that area.

THE POLLUTION STANDARD INDEX

Due to the wide range of contaminants, concentrations, and averaging times for the NAAQSs, the federal Environmental Protection Agency

Table 3.4. Nonattainment Areas for Criteria Pollutants, August 1992

Pollutant	Number of Nonattainment Areas[a]
Carbon monoxide (CO)	42
Lead (Pb)	12
Nitrogen dioxide (NO_2)	1
Ozone (O_3)	97
Particulate matter (PM10)	70
Sulfur dioxide (SO_2)	50

[a]Unclassified areas are not included in the totals.

Table 3.5. Pollutant Standard Index (PSI) Levels

PSI Range	Air Quality
0–50	Good
51–100	Moderate
101–200	Unhealthful[a]
201–300	Very Unhealthful[a]
>300	Hazardous[a]

[a]EPA designation.

(EPA) has determined that a normalized reporting system would be appropriate for public use. The *pollutant standards index* (PSI) has found widespread use in the air pollution field to report daily air quality to the general public. The PSI integrates information for the different criteria pollutants across an entire monitoring network into one normalized figure. This index, ranging in numerical values from 0 to 500, is intended to represent the worst daily air quality experience in an area.

The PSI is computed daily for particulate matter (PM10), ozone, carbon monoxide, sulfur dioxide and NO_2, based on *short-term* national ambient air quality standards, federal episode criteria, and federal significant harm levels. Lead is not included because it does not have a short-term air quality standard episode criteria or significant harm level.

The PSI converts daily monitoring information into a single measure of air quality by first computing a separate subindex for each pollutant and averaging time for that day. The PSI itself is a single dimensionless number. Table 3.5 lists the PSI ranges and their air quality designations for each range. It is used primarily to report daily air quality in a large urban area as a single number or descriptive word. Frequently the index is reported as a regular feature on the news media or in the newspapers.

It should be noted that in general, ozone and carbon monoxide receive the majority of the emphasis in the designation of PSI levels, since those are the two contaminants which cause most of the ambient air quality standard violations in urban areas. A PSI of zero is assigned to the limit of detection for the contaminant, and 100 is designated as the ambient air quality standard for whatever contaminant has the worst air quality in a given day.

Episodic Standards

There are concentrations of the gaseous criteria contaminants which will result in acute health impact. These episodic standards range from the

Table 3.6. Acute Health Impact Designations

PSI	Designation	Criteria Pollutant	Time, hr.	ppm
100	Federal AAQS	Ozone	1	0.12
		CO	1	35
200	Federal Alert	Ozone	1	0.20
	California Stage 1 Episode	Ozone	1	0.20
		CO	1	40
			12	20
275	California Stage 2 Episode	Ozone	1	0.35
		CO	1	75
			12	35
300	Federal Warning	Ozone	1	0.40
400	Federal Emergency	Ozone	1	0.50
500	Significant Harm Level	Ozone	1	0.60

Source: 40 CFR 51. Appendix L and Title 26, California Code of Regulations.

ambient air quality standards to levels of school health advisories to significant harm levels. Table 3.6 shows the acute health impact designations for two contaminants, ozone and carbon monoxide, which are the most likely to cause acute health impacts. They are listed by their respective PSI numbers.

The lowest designation is the federal primary AAQS with a PSI of 100. The respective one-hour concentrations are 0.12 ppm for ozone and 35 ppm for carbon monoxide. Levels in excess of 200 are a federal alert level and are designated as Stage 1 episode levels in California. This is the first level of warning for the general population where susceptible persons, especially those with heart or lung diseases, should stay indoors. Generally, healthy adults and children are strongly recommended to avoid vigorous outdoor exercise when these levels are exceeded.

When PSI levels of 275 to 300 (Stage 2 and Warning levels) are attained, the outdoor air quality environment has become hazardous. Therefore, all susceptible persons are directed to stay indoors, and the general population is recommended to avoid all outdoor activities. When PSI levels exceed 400, a federal emergency results. At PSI 500, significant harm results, and all individuals are to remain indoors and minimize physical activity.

Other regulatory actions take effect at these various stages in attempts to abate the acute health impacts either predicted or attained by these contaminants. Local agencies, of course, may designate their own warning or advisory levels to the population at large and to sensitive population groups, such as school children. For instance, within Southern California, a PSI of

138, which corresponds to an ozone concentration of 0.15 ppm, becomes a *school health advisory level* where prolonged vigorous outdoor exercise is not recommended.

NONCRITERIA AIR CONTAMINANT STANDARDS

There are two major approaches to determining acceptable air quality levels for noncriteria contaminants. The first of these is based on human health effects and extrapolations from epidemiologic data. This approach includes occupational exposure levels corrected by some factor less than 1.00.

The other approach is a health risk assessment whereby individual and the societal risks to exposed populations are factored into the determination of acceptable air quality concentrations. These are based on risk probability, and assume that carcinogens, mutagens, and teratogens, present in *any* amount, pose a risk.

Human Health Effects Based Standards

In this approach, regulatory agencies adopt acceptable concentrations in the ambient atmosphere of noncriteria pollutants. In general, these are based on health effects data either generated through toxicological studies or published data from the American Conference of Governmental Industrial Hygienists (ACGIH) in the threshold limit values (TLVs) or recommendations from NIOSH. These approaches modify occupationally-based levels acceptable in work situations to acceptable air concentrations for the general public. These approaches typically use the eight-hour TLVs and divide them by an appropriate factor such as 42, 100, 300, or 1000 to arrive at an acceptable concentration for the noncriteria air contaminants.

Table 3.7 lists a representative sample of acceptable ambient concentration guidelines for arsenic and its compounds for several states. Again, as with ambient air quality, the averaging times are different, as well as the numerical values between different states. It is interesting to note that only some of these standards are based on occupational exposure levels. In addition, some do and some do not have uncertainty factors listed for such concentrations.

A similar pattern is seen for formaldehyde in Table 3.8. Again, we see a

Table 3.7. Acceptable Ambient Air Concentration Guidelines or
Standards—Arsenic and Compounds

Agency	Concentration ($\mu g/m^3$)	Time, Hr.	OEL	Uncertainty Factor
Arizona	0.32	1	*	—
	0.084	24	*	—
	0.00023	Annual	*	—
Connecticut	0.05	8	*	200
Louisiana	0.02	Annual		—
Montana	0.39	24		—
	0.07	Annual		—
New York	0.67	Annual	*	300
Oklahoma	0.02	24	*	100
Rhode Island	0.0002	Annual		—
South Carolina	1.00	24	*	200
Texas	5.00	0.5		—
	0.5	Annual		—
Virginia	3.3	24	*	60

* = Based on an Occupational Exposure Level (OEL).
— = Not provided in database.
Source: NATICH Database Report of Federal, State, and Local Air Toxics
Activities, EPA 453/R-92-008, September 1992.

wide range of averaging times and concentration levels. There are a variety
of uncertainty factors and again only some of the levels are associated with
an occupational exposure level.

RISK ASSESSMENTS

For other hazardous air pollutants, a different approach is taken. This
approach focuses on the probability of an individual or a number of persons potentially contracting a fatal condition or being severely impacted by
an air contaminant.

When a risk assessment is performed, it is an attempt to derive a quantitative value of either a societal or individual risk. A societal risk is the *number*
of adverse consequences that might occur as a result of exposure to a
hazardous air pollutant. An individual risk is the *probability* of an adverse
consequence occurring to an individual during a year.

For health risk assessments, one is typically concerned with the potential
for a particular activity to cause chronic health problems. Principally, these
assessments deal with cancer rather than acute incidents such as injury or

Table 3.8. Acceptable Ambient Air Concentration Guidelines or
Standards—Formaldehyde

Agency	Concentration ($\mu g/m^3$)	Time, Hr.	OEL	Uncertainty Factor
Arizona	20.0	1	*	—
Connecticut	12.00	8	*	100
Indiana	6.00	8	*	200
Kansas	0.077	Annual	*	420
Louisiana	7.69	Annual		—
Massachusetts	0.33	24		—
Maine	67.0	0.25		—
	0.04	Annual		—
North Carolina	0.00015	0.25	*	10
Nevada	0.00007	8	*	42
New York	5.00	Annual	*	300
South Carolina	7.5	24	*	200
Texas	15.0	0.5		—
	1.5	Annual		—
Virginia	12.0	24	*	100
Vermont	0.08	Annual		—

* = Based on an Occupational Exposure Level (OEL).
— = Not provided in database.
Source: NATICH Database Report of Federal, State, and Local Air Toxics
Activities, EPA 453/R-92-008, September 1992.

death. Unfortunately, for a number of chemical contaminants, one is faced
with a relatively small body of health data. Therefore, many assumptions
and uncertainties are present.

The criteria air pollutants are not subject to a risk assessment; when these
standards were set, appropriate safety factors were included in setting such
standards.

The EPA has published formal guidelines to be followed for risk assessment. These include guidelines for carcinogens, mutagens, chemical mixtures and suspected developmental toxicants, and one guideline exists for
the estimation of exposure. As policy, the EPA holds that there is no safe
threshold concentration for a chemical carcinogen. The four major elements in a health risk assessment process, per the EPA guidelines, are:

1. hazard identification
2. dose-response assessment
3. exposure assessment
4. risk characterization

Hazard Identification

The hazard identification part of health risk assessment is a qualitative assessment based upon a review of relevant biological and chemical information. The elements involved in making a hazard identification of a given chemical are:

- physical and chemical properties of the agent
- routes of exposure
- structural or activity relationships that may support or argue against prediction of carcinogenicity
- metabolic properties of the agent
- toxicological effects of the agent
- information on both short- and long-term animal studies that have been performed to date

Since animal studies are frequently used in determining carcinogenicity in humans, these receive a great deal of weight in the hazard identification process. The weight of evidence that an agent is potentially carcinogenic for humans increases with:

1. the number of tissue sites affected by an agent
2. the number of animal species, strains or sexes that will show a carcinogenic response
3. the occurrence of clear-cut dose-response relationships in treated (as compared to control) groups
4. a dose-related shortening of the latency period with respect to the agent
5. a dose-related increase in the proportion of tumors that become malignant or true cancers

Data are classified alphabetically by groups (A through E) in descending order of overall weight of evidence:

Group A — data that relate to known carcinogenic effect to humans
Group B — probably carcinogenic to humans
Group C — possibly carcinogenic to humans
Group D — not classified as to human carcinogenicity
Group E — evidence of noncarcinogenicity for humans

To evaluate carcinogenicity, the primary comparison is tumor response in dosed animals as compared to other controls. Specific organs or tissues are also to be included in this identification.

Dose-Response Assessment

As noted earlier, the dose-response assessment is a critical element in determining the effect of a particular agent on carcinogenicity. The degree to which the agent is potent is important, as is its level of potency for a given dose, since these may vary widely among animal species and between animal species and humans.

Low Dose Extrapolations

The choice of the mathematical extrapolation model to very low doses must be made carefully. Risks at very low exposure levels cannot be measured directly by either animal experiments or by epidemiological studies. Therefore, mathematical models are used to try to fit the observed data. Extrapolations may be made to doses lower than presented in animal studies.

No single mathematical procedure is recognized as the most appropriate for low dose extrapolation in carcinogenicity. When data and information are limited and when much uncertainty exists regarding the mechanism of carcinogenic activity, models or procedures that incorporate low dose linearity are preferred when compatible with limited information.

The true value of the risk is *unknown* and may be as low as zero! Therefore, the range of risk defined by the upper limit given by the chosen model and the lower limit (which may be as low as zero) should be explicitly stated in the approach.

Comparison

Low dose risk estimates derived from lab animal data extrapolated to humans are complicated by a variety of factors that differ among species, and potentially affect the response to carcinogens. These items include life span, sex, body size, genetic variability, existence of other disease, exposure regimen, population homogeneity, and factors such as metabolism and excretion patterns.

Exposure Assessment

In order to obtain a quantitative estimate of carcinogen risk, the result of the dose-response assessment must be combined with an estimate of exposures to which the population of interest is likely to be subject. Table 3.9 outlines the major elements required when performing an exposure assessment.

The measured or estimated concentrations to be used for the exposure assessment will include the types of measurements, how they are generated, and an estimation of the environmental concentrations that could be expected. This may involve some form of mathematical modeling of emissions from either a point source, such as a stack, or an area source, such as a storage pile or landfill. Calculations of exposed populations, intake, and the uncertainties involved in the preceding calculations round out the exposure assessment. Background concentrations must also be presented and taken into account.

Table 3.9. Major Elements for an Exposure Assessment

1. Identification of Receptors and Routes of Exposure
   a. Human Populations at Risk
   b. Sensitive Subpopulations
   c. Locations and Habits
   d. Sources of Food and Water
2. Dispersion/Transport Modeling
   a. Environmental Fate of Suspect Chemicals
   b. Exposure Pathways—Air, Water, etc.
3. Estimation of Air Concentrations at Exposure Points
   a. Maximal Exposure Points
   b. Averaging Times
4. Chemical Intake Calculations
   a. Ambient Air—Inhalation
   b. Food and Water—Ingestion
   c. Dermal Contact
5. Pharmacokinetic Evaluation
   a. Absorption—Target Organs
   b. Metabolism
   c. Excretion
6. Estimation of Dose at Critical Organs

Source: U.S. EPA, Risk Assessment Guidance for Superfund, Vol I, Human Health Evaluation Manual, Part A (RAGS)—Interim Final (EPA/540/1-89/002).

Risk Characterization

Characterization of risk is composed of two parts: (1) a presentation of the numerical estimates of risk, and (2) a framework to help judge the significance of that risk. This framework consists of the exposure assessment and the dose-response assessment, and may also include a unit risk estimate. The latter can be combined with the exposure assessment for the purposes of estimating the cancer risk.

The numerical risk estimates can be presented in one of three ways:

- unit risk
- dose corresponding to given risk
- individual probability

Unit risk is the excess lifetime cancer risk attributable to a continuous lifetime exposure to "one unit" of carcinogen concentration under an assumption of low dose linearity. The second estimate is presented as the dose corresponding to a given level of risk (i.e., the dose corresponding to one excess cancer per million population exposed). This approach is used for nonlinear extrapolation models where the unit risk could differ at different dose levels. The third estimate is based upon individuals or populations (i.e., the probability of an individual or a group getting cancer for a given scenario).

In characterizing the risk due to concurrent exposure to several carcinogens, the risks are to be combined on the basis of additivity unless there is specific information to the contrary.

With respect to the latter, it should be noted that populations at particular risk include the elderly, the young, and those with preexisting health problems. Therefore, uncertainties on how they would respond to a given carcinogen or dose are critical compared to the general population at large.

UNCERTAINTIES IN RISK ASSESSMENTS

There are two major categories of uncertainties: the toxicity of a chemical agent for humans, and the exposure of a population group to that potential agent.

Uncertainties in Toxicity

Some uncertainties involve the testing scheme that was used in generating toxicity data on a given chemical for a given species. Others involve the potential synergism or antagonism that may occur between the chemical being observed and other coexisting chemicals or agents. It should be noted that tests are not an absolute measure of toxicity.

Whether or not a threshold can be presumed to exist is also important. The EPA considers carcinogens to have no acceptable threshold of exposure. If such a threshold exists, it would indicate that the organism can detoxify some level of chemical exposure (which is the case). If no threshold is assumed, this would indicate that some risk will always exist for any exposure to any level of a given chemical.

Extrapolations from high dose effects to low dose effects are another major area of uncertainty, with errors potentially ranging from 10^5 to 10^6 times. Interspecies differences and conversions of animal test data to humans are other sources of uncertainty involved in toxicity measurements. These two together will give uncertainties in the range of "orders of magnitude."

Uncertainties in Modeled Exposures

Much of the data used for risk management decisions on proposed projects potentially emitting toxic or hazardous substances involves some form of mathematical modeling. There are therefore serious concerns as to the degree to which a mathematical model is able to accurately determine the concentrations to which one may be exposed.

Typically, mathematical dispersion models for air pollution are accurate to within a factor of ±50–100%. Emission estimates (source terms) for point sources of criteria pollutants are generally accurate to within ±20%.

SCREENING LEVEL APPROACHES

While the foregoing indicate the steps in an exhaustive risk assessment, there are a number of circumstances in which a *screening level* approach is considered a viable alternative. The screening approach eliminates the extreme cost and time of an exhaustive health risk assessment. For noncarcinogens, the *Hazard Index* is evaluated.

For each contaminant, a number of acceptable exposure levels (AELs)

Table 3.10. Typical Noncancer Acceptable Exposure Levels (Chronic)

Substance	Inhalation[a] ($\mu g/m^3$)
Ammonia	1.0E+2
Arsenic	5.0E-1
Benzene	7.1E+1
Cadmium	{3.5E+0}
Carbon tetrachloride	{2.4E+0}
Chlorinated dibenzo-p-dioxins (as 2,3,7,8-equivalents)	{3.5E-6}
Chromium (hexavalent)	2.0E-3
Ethylene oxide	6.0E+2
Formaldehyde	3.6E+0
Gasoline vapors	2.1E+3
Hydrogen cyanide	{7.0E+1}
Mercury and compounds (inorganic)	3.0E-1
Methanol	6.2E+2
Methylene chloride	3.0E+3
Perchloroethylene	{3.5E+1}
PCBs (polychlorinated biphenyls)	1.2E+0
Toluene	2.0E+2

[a]Values in { } have been converted from oral acceptable exposure levels (mg/kg/day) by assuming a 70-kg person breathes 20 m^3 per day and equal absorption occurred by the inhalation and oral routes. E denotes scientific notation, base 10.

have been adopted, as noted earlier. These include chronic noncancer (Table 3.10) and acute noncancer (Table 3.11) acceptable exposure levels.

An individual hazard index is computed by dividing a mathematically modeled concentration by the appropriate AEL. This hazard index is performed for each of the contaminants of concern. If the hazard index is less than 1.0, it is considered acceptable for the exposed population. The individual hazard indices may be totaled over all contaminants to give a total hazard index for a specific toxicological endpoint, such as respiratory impact. If the hazard index exceeds 1.0, it is a cause for concern as to the cancer impacts and is subject to a risk management decision.

Carcinogen Hazards

For carcinogenic compounds, a slightly different approach is taken. In this approach a "unit risk factor" is utilized. The *unit risk factor* (URF) is a

Table 3.11. Typical Noncancer Acceptable Exposure Levels (Acute)

Substance	Exposure Level ($\mu g/m^3$)
Ammonia	2.1E + 3
Carbon tetrachloride	1.9E + 2
Formaldehyde	3.7E + 2
Hydrogen cyanide	3.3E + 3
Hydrogen sulfide	4.2E + 1
Mercury (inorganic)	3.0E + 1
Methyl chloroform	1.9E + 5
Nickel compounds	1.0E + 0
Perchloroethylene (tetrachloroethylene)	6.8E + 3
Xylenes	4.4E + 3

E denotes scientific notation, base 10.

summation of all of the factors from a rigorous health risk assessment into one number, which combines absorption, target organ, conservative assumptions, most responsive individual, etc. The unit risk factor, therefore, is the *individual probability* of contracting cancer when one is exposed over a 70-year lifetime to an ambient concentration of one microgram per cubic meter of a given carcinogen. The unit risk, when multiplied by the dispersion modeled concentration, yields the individual risk. Table 3.12 lists the URFs for a number of compounds.

Thus, if the modeled concentration of formaldehyde is 5 micrograms (μg)/m^3, the individual risk would be:

$$\text{Concentration} \times \text{URF} = \text{individual probability}$$

or

$$5 \ \mu g/m^3 \times \frac{1.3 \times 10^{-5}}{\mu g/m^3} = 6.5 \times 10^{-5}$$

If the modeled concentration of formaldehyde was 0.1 $\mu g/m^3$, then the corresponding cancer risk would be 1.3×10^{-6} which is slightly above the typical acceptable cancer risk of 1×10^{-6}.

Thus for cancer, the screening level assessment approach allows one to take the unit risk factor and multiply it times the maximum concentration (derived by air dispersion modeling) to determine the individual cancer risk.

Table 3.12. Typical Cancer Potency Values

Substance	Unit Risk Levels ($\mu g/m^3$)
Arsenic	3.3E–3
Benzene	2.9E–5
Cadmium	4.2E–3
Carbon tetrachloride	4.2E–5
Chromium (hexavalent)	1.4E–1
Formaldehyde	1.3E–5
Methylene chloride (dichloromethane)	1.0E–6
Nickel and nickel compounds	2.6E–4
PCBs (polychlorinated biphenyls)	1.4E–3
Perchloroethylene (tetrachloroethylene)	5.8E–7
Trichlorethylene	2.0E–6
Vinyl chloride	7.8E–5

E denotes scientific notation, base 10.

The actual number of individuals impacted (societal risk) are those living within a geographic area encompassed by a given cancer risk such as 1 chance in one million (10^{-6}).

COMPARISON OF AIR QUALITY TO STANDARDS

For the criteria pollutants, there are a number of ways of expressing whether ambient air quality is within the acceptable range. These include the number of "exposure hours," the number of "station days," or percent of time above a standard. For purposes of the Clean Air Act, the most important value for determining *attainment* of the NAAQS is the second highest concentration. This becomes the design value for air quality management strategies. The number of days that any ambient air quality standard was exceeded in a year indicates the severity of the problem.

Figure 3.3 indicates the second highest daily maximum one-hour average ozone concentration in 1991 for the continental United States. In this figure, we see that the maximum hourly ozone concentration was slightly above 0.3 ppm in Los Angeles, with a number of other locations exceeding the ambient air quality standard of 0.12 ppm. Figure 3.4 compares the second highest maximum *daily* average particulate concentration for differ-

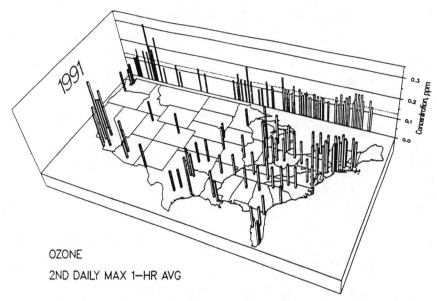

OZONE
2ND DAILY MAX 1—HR AVG

Figure 3.3.   United States map of the highest second daily maximum 1-hour average ozone concentration by MSA, 1991.

ent geographic areas of the country for metropolitan areas with a population of greater than 500,000. Six metropolitan areas exceeded the ambient air quality standard of 150 micrograms (μg)/m^3.

MONITORING AIR QUALITY

Monitoring of air quality is important for three major reasons. The first of these is to survey an area to establish contaminant concentrations in the immediate vicinity of a "hot spot" of emissions. This could be near a suspected source where the concentration is high due to very little dispersion or a release close to the ground. The second major concern is with determining the air quality in a given location for health advisories to sensitive populations, athletes, the elderly, schoolchildren, etc. The third major concern is to determine compliance with federal law regarding the attainment of the national ambient air quality standards.

The location criteria for an air monitoring station is therefore different for each of the above purposes. For all three, a secure location with adequate power, security, and protection from the elements are among the

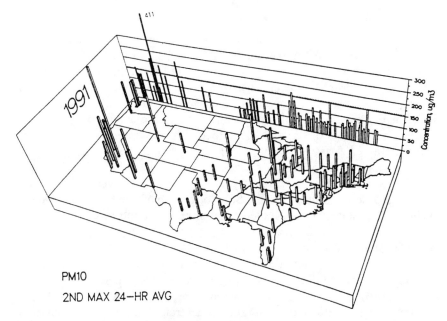

Figure 3.4.   United States map of the highest second maximum 24-hour average PM-10 concentration by MSA, 1991.

common requirements. Other distinctions arise, however, depending on which purpose is being pursued by an air monitoring agency.

In general, *hot spots monitoring* involves a short time period, usually a few days to a few months, but includes around-the-clock air monitoring. The concerns here are for acute concentrations due to potentially catastrophic releases, as well as concerns for chronic health impacts due to high average concentrations. These monitoring locations tend to be at the property line of a suspected hot spot source. Monitoring methods for hot spots tend to be nonstandard in their location criteria.

For air monitoring where health advisories may be the prime concern, i.e., schools, hospitals, etc., the duration is typically short-term or seasonal. Such a case would be for ozone during the summer in an urban area. Consequently, such installations would tend to be of short duration during certain periods of the day. Like hot spots monitoring, these monitoring installations are set up for only one or two key pollutants. These installations may be fully automated or battery-powered, with little human contact. Telemetry systems, sending information directly to a central location, or strip-chart recorders recording data for two to four days without human intervention, are considered normal.

For the purpose of determining attainment of ambient air quality standards, an entirely different picture arises. The purposes for such an installation are to gather representative air quality data for the entire metropolitan statistical area (MSA) to determine compliance with the Clean Air Act and other provisions of law. Therefore, these installations tend to be in fixed locations for many years at a time, and tend to use continuous analyzers housed in buildings or trailers. A number of analyzers are usually present and maintained at a given location. The purposes of these stations are not only to monitor air quality for seasonable variations and acute episodic conditions, but also to determine the average air quality over a long period of time in order to evaluate trends. Since operation of these stations tends to be by local or state agencies under the authority (and meeting the guideline requirements) of the Environmental Protection Agency, they are termed SLAMS (state and local air monitoring stations).

The stations established to monitor national air quality trends are called the National Air Monitoring Station (NAMS) sites. These are generally operated by or with a cooperative agreement with the EPA. The NAMS are located in areas with high pollutant concentrations and high population exposure. These stations meet uniform criteria for siting quality assurance, equivalent analytical methodology, sampling intervals, and instrument selection to ensure consistent data reporting among the different sites.

Other sites operated by the state and local air pollution control authorities, such as special purpose monitors (SPM), in general meet the same criteria as NAMS, except that they are located in other areas as well.

Prevention of Significant Deterioration (PSD) stations are privately funded and operated air monitoring stations set up to specifically meet the PSD requirements of the Clean Air Act. They generally are operated for one full year before and after installation of large stationary sources and have specific requirements for air monitoring equipment.

Measurement Techniques

Early methods to measure ambient levels of contaminants were limited to observations of their effects. As an example, ozone was measured over a period of several hours using stretched rubber bands in Los Angeles, since it is known that ozone was responsible for rubber cracking. Particulate matter was "measured" primarily at airports in terms of the visible range from the airport to fixed objects at different distances.

These techniques did give some indication of the level of air contamination, and usually included features such as an *eye irritation index.* They did not provide real-time information, since they were simply an observation noticed in passing.

Cumulative Samplers

The earliest numerical methods were simple weight-based measurements of particles as "dustfall." This measured the weight of heavy particles falling into a glass jar on a monthly basis. Sulfur dioxide was indirectly monitored using lead-based pastes coated on glass cylinders. The ambient SO_2 reacted with the media and formed an insoluble lead sulfate which was later analyzed in a chemistry laboratory. This yielded the "flux" of SO_2 per unit area on the surface of the sampler.

Other early techniques to measure average concentrations were extractive samplers. These methods used absorption solutions through which was drawn a known volume of ambient air in a glass bubbler. These solutions were analyzed in a chemical laboratory, and calculations were made as to determine the concentration after the fact.

Integrated or cumulative samplers are used today for determining some criteria pollutant concentrations. The reference method for SO_2 is a wet chemical absorption technique followed by analysis in a laboratory. Such a gas sampling train is seen in Figure 3.5. Hydrogen sulfide (in California) is also measured with the reference technique using a bubbler solution and chemical indicator followed by chemical laboratory analysis.

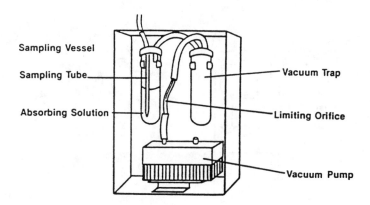

Figure 3.5.   A gas sampling train.

Particulates are also measured by using an integrated or cumulative sampler, called a high volume sampler (Hi Vol). It consists of a special housing mounted on top of a pre-weighed particulate glass fiber filter. Mounted below is a 30 cubic foot per minute blower and motor. In this technique, air is drawn through the filter at a known rate until the end of a 24-hour sampling period. Following that, the filter is taken off the unit, dried, and weighed in the laboratory.

The volume is calculated based upon the unit flow rate. The determination of PM10 concentrations requires the installation of a size-selective inlet head on top of the filter chamber. The size-selective inlet takes advantage of inertia to separate out the larger particles (greater than 10 microns) from the fine particles which are collected on the backup filter. From the mass, the concentration is calculated. Total particulate may be determined in the same manner, only the special inlet is not used. Figure 3.6 illustrates a HiVol sampler.

Lead is determined by analyzing a portion of the same filter catch and calculating the concentration of lead in that same sample. Other analyses may be performed on that filter to determine proportional concentrations of other contaminants such as sulfates, nitrates, etc.

Particulate may also be measured using an automated discrete tape sampler which draws a one- to two-hour sample of ambient air through a paper tape. The optical reflectivity and density of the now-colored spot are measured, and the coefficient of haze (COH) is printed out on a recorder. The tape then advances to a new portion of the paper, and sampling resumes.

For hot spot sampling, as well as special studies, integrated or cumulative samplers are also used. These primarily involve either a wet chemical bub-

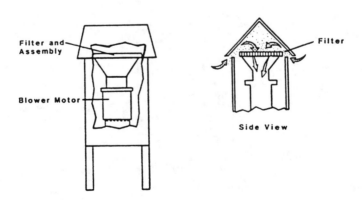

Figure 3.6.   HiVol sampler.

bler system, as seen above, or using evacuated polished stainless steel cylinders or spheres into which ambient air is slowly drawn over a certain period of time. At the close of the sampling period, these containers are taken to the laboratory and analyzed for the contaminants of interest. The latter are used for determining air toxic (primarily organic gas) concentrations at very low concentrations in ambient air.

Some of the techniques, such as for ozone and NO_2, were measured using automated "wet" chemical analyzers. In these approaches the measurement was made in the analyzer with a readout of the contaminant concentration within several minutes of the sample passing through the instrument.

In general, the techniques for monitoring air quality at the present time have moved away from cumulative samplers and tend to use continuous analyzers. Standard reference methods, or those proven equivalent to them, are required for official air monitoring purposes.

Continuous Analyzers

It is understood that while reference methods may call for wet chemical-based systems, it is possible to utilize instrumental techniques that have been shown to give results equivalent to the reference methods.

For the criteria pollutants, ozone, carbon monoxide, and nitrogen dioxide, continuous analyzers are used for determining ambient concentrations. These electronic analyzers continuously take samples of ambient air and pass it through a detector, which gives a signal proportional to the concentration. The microprocessor in most of these analyzers stores the concentration for later use.

Each of the continuous analyzers operates on a different principle, but is primarily based on the response of a gaseous pollutant to an electromagnetic property. The light emission generated in the analyzer is directly proportioned to the pollutant gas concentration.

Ozone, for instance, can be measured by ultraviolet photometry or by chemiluminescence. Ultraviolet photometry operates by measuring the light emission of ozone at a specific wave length under an initial impulse of radiation.

Both ozone and nitrogen dioxide, as measured by chemiluminescence or "gas phase titration," operate by injecting the corresponding opposite gas into the analyzer and measuring the light emission at a specific wave length as NO_2 reacts with ozone or as ozone reacts with NO_2. Each requires a different instrument and a different wavelength for the measurement of the contaminant.

Carbon monoxide operates by a nondispersive infrared spectrophotometric technique. The carbon monoxide in the air sample absorbs heat in the analyzer as it is passing through the detection unit. Heat absorption causes a small diaphragm to move by differential pressure, which gives an analog response proportional to the carbon monoxide concentration. Other instrumental analyzers operate by measuring some optical or electronic signal generated by the gas.

Total hydrocarbons are typically measured by a flame ionization system. Here the ionized carbon atoms in an ambient gas sample move across an impressed voltage in the detector and are collected on a plate. This generates an electrical signal proportional to the carbon atom concentration in the gas sample.

Other techniques which may be used include semi-continuous analyzers such as gas chromatographs which take samples at regular intervals and analyze them as if they were laboratory samples. These field instruments are able to give a speciation of different contaminants as well as their concentrations.

Some hot spot analyses or survey instruments used in determining noncriteria air pollutants involve taking samples over a very short duration of time in a container, such as a Tedlar bag with a nozzle or an evacuated glass flask. These samplers, when filled, are sent to a laboratory for analysis.

4 SOURCES AND MEASUREMENT METHODOLOGIES

In the city, because of . . . all that pours forth from its inhabitants and their superfluities . . . the air becomes stagnant, turbid, thick, misty, and foggy . . . we have grown up in the cities and have become accustomed to them. . . . Moses Maimonides,
Cairo, ca. 1200

If society is to properly manage its air quality resources it must understand fully the sources and distribution of the various air contaminants. Air contaminants may be emitted to the atmosphere from three major source groups: geogenic, biogenic, and anthropogenic sources. The first two refer to essentially natural air pollution, whereas the latter refers to the activities of mankind. It is important to keep the relative contributions of each category in mind in our air quality management strategy development.

Certain of these sources have the air quality advantage of being widely dispersed around the earth's surface. Only 21% of the earth's surface is solid land and of that, only a fraction has population densities sufficient to experience the air pollution impacts of man's activities.

Geogenic Sources

Volcanoes, lightning, emissions from geothermal springs, wildfires, ocean floor "smokers," and dust storms are all sources to which, by comparison, mankind's sources seem trivial. Volcanoes are noted for their acute

77

impacts, primarily by virtue of deaths in the immediate vicinity of the volcano itself. The eruption of Mt. Tambora (Indonesia) in 1815 claimed 92,000 lives. Krakatoa, in 1883, in the Indonesian Islands, drowned 36,000 people with a tsunami, and the blast was heard 3,000 miles away. Mount St. Helens, a rather average volcano, on May 18, 1980, killed 60 people when it blew off one cubic kilometer of material in one day.

There are approximately 500 known active volcanoes on the earth and upward of 500 million people living in close proximity to them. Some estimates indicate that between 10 and 25 volcanoes are erupting at any given moment in time.

In addition to immediate disasters and acute loss of life, the material blasted into the sky has been noted to have long-term global impacts as well. The 1815 eruption of Tambora put so much material into the atmosphere, that the cloud of dust and gases lowered temperatures around the world. In the United States, 1816 was known as the "year without a summer." Temperatures were so low that snow actually fell during the summer.

If one reviews the amount of material blown into the atmosphere by a volcano such as Mount St. Helens, it is possible to calculate the approximate tonnages of the elements which become air pollution. Table 4.1 is a listing of the estimated tonnage of the various elements which become airborne in one day from the Mount St. Helens blast of May 18, 1980. In addition, the volcano was noted to have several eruptions prior to that date and for a year thereafter. These figures, therefore, represent only one day's air emissions for one average-sized volcano.

Table 4.1. Mount St. Helens Air Emissions

Element	Release (Tons)[a]
Mercury	235
Cadmium	590
Arsenic	5,300
Lead	35,000
Zinc	206,000
Chromium	294,000
Chlorine	382,000
Sulfur	764,000
Barium	1,250,000

[a]Calculated from average density of 2.9 tons/cubic meter and *Handbook of Chemistry and Physics,* 72nd ed., CRC Press, Inc., Boca Raton, FL, 1992.

Biogenic Sources

Biogenic sources of emissions include those from living organisms and the decay of formerly living organisms. With respect to the former, we find plants and crops as well as trees, shrubs, and grasses. Air emissions also occur from bacteria, protistans, and other phytoplankton in the oceans. They, likewise, contribute to the global picture of air contaminant emissions. These form a part of the natural carbon and nitrogen cycles, without which there would be no living matter. Decaying organic matter also contributes to air emissions by way of soil and water. Emissions also occur naturally from fires and landfills.

Anthropogenic Sources

Anthropogenic air emissions result from fuel combustion, evaporation of a variety of materials, and by direct emissions from activities generating particulates or aerosols. These emissions are differentiated from those of the indoor environment, which have much different source characteristics.

GLOBAL SOURCE COMPARISONS

Table 4.2 summarizes estimates of global air emissions in millions of metric tons per year of various criteria air contaminants. Also shown is a breakdown of the amounts that are attributable to natural and to anthropogenic causes. For each contaminant, the total mass of nature's emissions are far in excess of those generated by man. Natural emissions are seen to range from 10 to more than 1,000 times the man-made sources.

If one looks at just one element, such as sulfur, which reacts to become

Table 4.2. Global Air Emissions (Millions of Metric Tons Per Year)

Source	CO	Total NO_x	HC	H_2S	PM10
Natural	3,000–643,000	477–990	115–1,800	100	(80–90%)
Anthropogenic	260–1,060	48–53	30–90	3	(10–20%)

Source: Legge, A. H., and S. V. Krupa, *Acidic Deposition: Sulphur and Nitrogen Oxides,* Lewis Publishers, Chelsea, MI, 1990; Calvert, S., and H. M. Englund, Eds., *Handbook of Air Pollution Technology,* John Wiley & Sons, New York, 1984; Finlayson-Pitts, and Pitts, *Atmospheric Chemistry: Fundamentals and Experimental Techniques,* John Wiley & Sons, New York, 1986.

sulfur dioxide and ultimately sulfates, one is impressed by the differences in source contributions. Table 4.3 indicates the tons per year (millions) estimated for a variety of sources including volcanoes, sea spray, ocean and land mass emissions from biogenic sources.

The proportions of global NO_x are seen in Figure 4.1. These indicate calculations of the percentages by source of the total contribution of global NO_x around the world, including anthropogenic sources. Researchers have computed different quantities, as in Table 4.4, for the global oxides of nitrogen emissions breakdown. In this table, the two primary species (N_2O and NO) are compared for natural and anthropogenic sources. In those computations, the man-made sources are less than five percent of the total worldwide emissions.

While not quantified, natural occurring emissions of nonmethane hydrocarbons, and in particular terpene compounds based upon the highly reactive five carbon *isoprene* unit are not only present in significant quantities

Table 4.3. Sulfur Budget — Global Emissions

Source	10^6 Metric Tons/Year
Natural:	
Sea spray	44
Biogenic (land)	48
Biogenic (ocean)	50
Volcanoes	5–100
Anthropogenic	104
Totals:	251–346

Source: Legge, A. H., and S. V. Krupa, *Acidic Deposition: Sulphur and Nitrogen Oxides,* Lewis Publishers, Chelsea, MI, 1990.

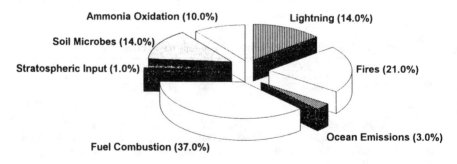

Figure 4.1.   Global NO_x contribution (percent by source).

**Table 4.4. Nitrogen Oxide Compounds,
Global Emissions Breakdown by Species**

Primary Species	10^6 Tons/Year	Source
N_2O	540	Natural
NO	450	Natural
NO	48	Anthropogenic

Source: Calvert, S., and H. M. Englund, Eds. *Handbook of Air Pollution Technology,* John Wiley & Sons, New York, 1984.

worldwide, but also tend to occur in those areas where people are most likely to be present. Such a case are the Great Smoky Mountains in the eastern United States. It is believed that emissions from soft wood forests produce sufficient quantities of natural hydrocarbons such as isoprene, combined with the existing levels of oxides of nitrogen, to produce natural concentrations of ambient ozone of the same order of magnitude as the national ambient air quality standard.

AIR POLLUTION SINKS

In addition to sources of air pollution globally and locally, there are also air pollution sinks. These "sinks" are processes by which air contaminants are consumed or removed from the atmosphere in sufficient quantities that, on a global basis, contaminant concentrations are in near equilibrium.

Biological Sinks

There are three major mechanisms by which air pollution is removed from the atmosphere. The first of these is by biological action. In this mechanism, air contaminants are absorbed into water, either in the atmosphere, in the oceans, or in soil moisture. Following that absorption, metabolic processes in single and multicellular animals and plants convert these contaminants to other compounds, where they become a part of the biological chain. Air contaminants are converted to cell mass and ultimately become an organic sludge or other simple gases. In the latter case they enter the atmosphere once again.

Carbon monoxide (CO), appears to be one of the key air pollutants which are converted into other metabolites by this mechanism. It has been estimated that the soil in the United States alone is estimated to have a CO removal *capacity* of 500 million tons per year. This is greatly in excess of the anthropogenic emission rate. To some degree, sulfur dioxide may be removed by virtue of absorption directly into the leaf tissues of plants.

Mechanical Sinks

A second major mechanism for removal of air pollution is by mechanical action. In this mechanism, large particles fall out directly from the atmosphere by gravity, or are absorbed in rainwater and washed out of the atmosphere. Aerosols are affected by further reactions in the atmosphere and subsequent removal processes.

Photochemical Sinks

The mechanism responsible for by far the greatest removal of air pollutants in the atmosphere are atmospheric processes themselves. These are primarily oxidation reactions and are photochemically induced. Table 4.5 illustrates sources and sinks for a number of contaminants and their estimated "half-lives" in the atmosphere.

In this mechanism, the various contaminants are subject to other reactions, principally oxidative. Sulfur dioxide is ultimately converted to sulfate ion, which is subject to the mechanical removal actions above. NO_x may be transformed to nitric acid and then nitrate, with subsequent removal by mechanical action. Or it may be further reacted in the ozone, producing a chain of chemical reactions. Nitrates may be further reacted in aerosols to form ammonium nitrate salts, which may be removed by either dry or wet deposition processes.

Carbon monoxide and hydrocarbons will react directly in the atmosphere with oxides of nitrogen and be removed by photochemical activity to form ozone. CO may be further reacted to carbon dioxide, and thus remain in the carbon cycle. Volatile organic compounds are oxidized to CO and CO_2, where they remain in the gaseous state pending further reactions and/or removals.

Oxidation products such as ozone and peroxyacyl nitrates (PAN) will remain until they are either reduced back to oxygen, further oxidized, or react directly with living tissues, fabrics, or materials.

Table 4.5. Gaseous Air Pollutants, Sources/Sinks/Residence Times

Pollutant	Anthropogenic Source	Natural Source	Rural Atmospheric Concentration	Residence Time	Principal Sinks
CH_4	Combustion	Biogenic	1.5 ppm	8 years	Reaction with OH
CO	Auto exhaust	Forest fires + photochem	0.25 ppm	1–3 months	Oxidation
CH_3Cl	Combustion	Oceanic	600 ppt	1–2 years	Stratospheric reactions
CO_2	Combustion	Biological	345 ppm	2–4 years	Biogenic
HCl, Cl_2	Combustion, manufacturing	Volcanoes	0.5 ppb	7 days	Precipitation
H_2S + Sulfides	Sewage + chemicals	Volcanoes + biogenics	0.15 ppb	1–2 days	Oxidation
NO, NO_2	Mobile & fossil fuel combustion	Biogenic & lightning	0.1 ppb	2–5 days	• Oxidation • Precipitation • Dry deposition
NH_3	Waste treatment	Biogenic	1–10 ppb	1–7 days	• Reaction with SO_2 • Oxidation
N_2O	Combustion	Biogenic	330 ppb	20–100 years	Photochemical in stratosphere
SO_2	Fossil fuel (combustion)	Volcanoes	0.05–20 ppb	2–4 days	Oxidation

ANTHROPOGENIC AIR EMISSIONS

The major anthropogenic emissions are grouped by the type of generating *mechanism:* combustion, evaporation, fugitive or waste related emissions. These latter include those from solid and hazardous waste treatment or disposal, or from remediation activities. Each of these will be detailed in the following sections.

Combustion

The three primary requirements for combustion to occur are seen in Figure 4.2, the combustion triangle. If any one of these three elements is removed, then combustion cannot occur. These elements are:

- air or oxygen
- fuel
- energy or heat

In general, we think of combustion as a fairly simple chemical reaction. In combustion, a fuel such as a hydrocarbon is reacted with oxygen in the presence of sufficient energy to begin the reaction. The ultimate results are the complete products of combustion. An example of these exothermic reactions are:

$$C + O_2 \rightarrow CO_2 + heat$$

$$H_2 + O_2 \rightarrow H_2O + heat$$

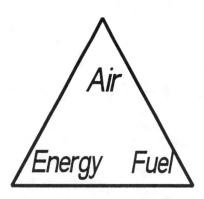

Figure 4.2.   The combustion triangle.

In the simple examples given above, the net result is full oxidation of the carbon and hydrogen, with corresponding products of water and CO_2. An excess amount of heat is released. This excess heat is the useful energy which we derive from the system to provide warmth or power.

Another way of looking at that process is to review Figure 4.3, in which an energy diagram is shown for a similar oxidative reaction. In this reaction, initially the original materials (methane + Cl) only collide with each other. At some point the energy of activation, ΔH^* (enthalpy), must be added to the system for the reactants to form an intermediate high energy state. The activation energy plus excess heat is then released as the products proceed to the oxidized states.

The resulting enthalpy (ΔH), is at a lower value than the initial reactants, and represents the useful energy which is released from the system. Should the input energy be less than required to attain the intermediate state's activation, the reaction will not go to completion and the original materials will remain. Or, if heat is withdrawn from the system prior to that activation energy being attained, again, the reactions will not occur. This represents a quenched reaction in which heat is removed prior to the completion of the overall reaction.

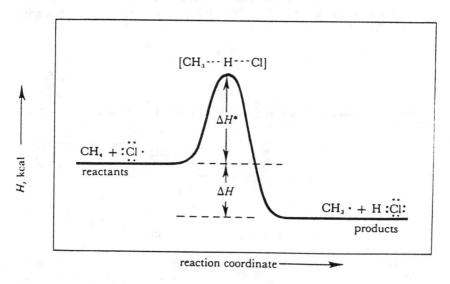

Figure 4.3.   Energy diagram for the reaction, $CH_4 + Cl\cdot \rightarrow CH_3\cdot + HCl$.

Fuels for Combustion Reactions

In the air pollution field, we are primarily dealing with carbon-based systems, i.e., fossil fuels. These range from the simplest methane molecule (CH_4) up to very complicated chemical structures such as coal.

Coal

Coal is one of the most common hydrocarbon fuels worldwide. In the United States, it supplies nearly 60% of all of the energy produced. Coals are relatively cheap, but have the disadvantage of high pollution potential.

Table 4.6 summarizes typical analyses of a variety of the more popular U.S. coals and lignite. One of the more interesting aspects is the wide range in sulfur content, and the relatively high ash content, which forms particulates. Generally, a higher total sulfur content in coal is accompanied by higher iron content. This is primarily due to iron and sulfur occurring in the form of pyrite, FeS_2.

A great deal of attention is placed on mineral analyses of coal ashes. When coal is burned in bulk form, most of the mineral content remains either as bottom ash or fly ash. These minerals were introduced in the coal by the original vegetation and by interdiffusion of the organic matter with the soil which bore the original plants.

Table 4.6. Ash and Mineral Content of Some U.S. Coals and Lignite

	Rank			
	Low Volatile Bituminous	High Volatile Bituminous		Lignite
	West Virginia	Illinois	Utah	Texas
Component				
Ash[a]	12.3	17.36	6.6	12.8
Sulfur[a]	0.7	4.17	0.5	1.1
Analyses, % of Ash				
SiO_2	60.0	47.52	48.0	41.8
Al_2O_3	30.0	17.87	11.5	13.6
Fe_2O_3	4.0	20.13	7.0	6.6

[a]Dry basis, % of fuel weight.
Source: Steam, Its Generation and Use, Babcock and Wilcox, New York, 1978.

The major constituents of the ash are oxides of silicon, aluminum, and iron with lesser amounts of titanium, phosphorus, sulfur, magnesium, and the alkali metals. These also will exhibit variations not only from region to region, but also significant variations will be observed within a particular region or even an individual coal seam.

In addition to the major ash constituents, numerous trace elements, most of which are HAPs, are also found in coal. Trace elements typically found in U.S. coal ash from all regions are beryllium, chromium, lead, manganese, mercury, nickel, and selenium. Also present on a less widespread basis is arsenic. It should be noted that with few exceptions, trace elements are more abundant in coal than they are in the earth's crust or soil. Also, levels of some elements, notably cadmium, mercury, and selenium are generally not reported. These elements are not normally detected in ash samples due to their higher volatility.

The importance of the ash comes not only in terms of air pollution, but also in terms of the properties of the deposits that occur. These ashes cause corrosion in the combustion equipment and catalytic effects, converting sulfur dioxide to sulfur trioxide/sulfuric acid. When emitted as large particulates, deposits may create a local nuisance. Also, deposits lower heat transfer, thereby forcing consumption of larger quantities of fuel. This indirectly contributes to greater combustion emissions.

Solid fuels such as coals or others such as bagasse (the cellulosic material left after the removal of sugar from sugar cane), petroleum coke, and wood, find various geographic niches in the combustion fuel picture. Concerns are similar to those for coal, but the nature of the ash formed is significantly different.

Liquid Fuels

Fuel oils are the residual materials left over from crude oil after the extraction of natural gas, gasoline, kerosene, and diesel fractions. They influence combustion emissions by the manner in which the fuel is burned, as well as by its composition.

Table 4.7 indicates typical analyses of fuel oils commonly burned in boilers or other combustion devices. Significant differences exist here. Most notable are the much lower sulfur contents and the trace quantities of ash in these oils as compared to coal. The lower quantities of ash and sulfur lead to higher heating values per pound of fuel oil. Whatever mineral matter existed in the fuel will also be oxidized and emitted in a particulate form. These may include trace volatile elements such as mercury, arsenic, and selenium.

Table 4.7. Range of Fuel Oils Analyses

Grade of Fuel Oil	No. 2	No. 4	No. 6
Analysis[a]			
Sulfur	0.05–1.0	0.2–2.0	0.7–3.5
Hydrogen	11.8–13.9	(10.6–13.0)[b]	(9.5–12.0)[b]
Carbon	86.1–88.2	(86.5–89.2)[b]	(86.5–90.2)[b]
Nitrogen	Nil–0.1	–	–
Ash	–	0–0.1	0.01–0.5
Heating Value			
Btu per lb, gross (calculated)	19,170–19,750	18,280–19,400	17,410–18,990

[a]Weight percent
[b]Estimated
Source: Steam, Its Generation and Use, Babcock and Wilcox, New York, 1978.

Distilled fuels for mobile power plants such as automobiles and airplanes utilize gasolines, kerosenes, and aviation fuel. These fuels form a compromise between the easy handling characteristics of fuel oils with the near zero sulfur content and lack of ash components of natural gas.

On the other hand, there are fractions of gasoline which have components which are considered hazardous, such as benzene and similar aromatic compounds. They pose a threat to public health based on evaporation tendencies. In addition, there are trace quantities of heavier molecular weight compounds which may have carcinogenic, corrosive, tar or deposit-forming characteristics.

One technique for evaluating the fractions of a liquid fuel which are of higher molecular weight is seen in Figure 4.4. This is a *distillation curve* which compares the fractions of liquid fuels which boil at various points. The T90 point (temperature at which 90% of the fuel has boiled off) forms somewhat of an index of the amount of heavy hydrocarbons present in each fuel. Methanol, for comparison, has a single boiling point in its pure state and has been considered as an alternative "clean" fluid due to its lack of major impurities.

Natural Gas

Table 4.8 summarizes the analyses of common natural gases found in the United States. With gas there is no ash. Also, being a gas, it is much easier

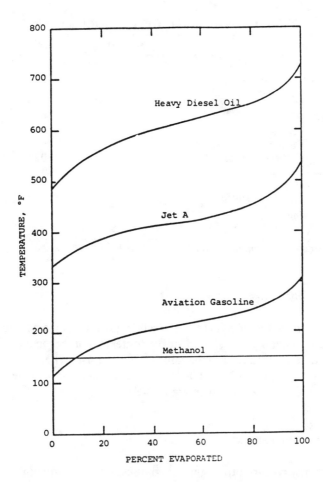

Figure 4.4.   Liquid fuel distillation curves.

to transport. The heating value per pound of natural gas is the highest of all of the fossil fuels. The rating is commonly given in terms of heat content (BTUs) per cubic foot of gas.

Natural gas is considered the most desirable fuel from an air pollution perspective. It is piped directly to the consumer, eliminating the need for storage at the consumer's plant. It is free of ash and mixes intimately with air to provide complete combustion at low excess air without smoking. The absence of sulfur (taken out during processing) makes it a highly desirable fuel.

Table 4.8. Natural Gas from United States Fields

		So. Cal.	Ohio	Okla.
Analyses				
Constituents, % by vol				
H_2	Hydrogen	—	1.82	—
CH_4	Methane	84.00	93.33	84.10
C_2H_4	Ethylene	—	0.25	—
C_2H_6	Ethane	14.80	—	6.70
CO	Carbon monoxide	—	0.45	—
CO_2	Carbon dioxide	0.70	0.22	0.80
N_2	Nitrogen	0.50	3.40	8.40
O_2	Oxygen	—	0.35	—
H_2S	Hydrogen sulfide	—	0.18	—
Higher Heating Value				
Btu/cu ft @ 60°F & 30 in. Hg		1,116	964	974

Source: Steam, Its Generation and Use, Babcock and Wilcox, New York, 1978.

Efficiency and Emissions

The heating value of a fuel is relatively important, since with lower grade fuels (containing higher ash and sulfur content), a larger mass of fuel must be burned in order to release the same amount of useful heat. Therefore, the total emissions per million BTUs will be greater for a lower grade, or "clean," fuel than for a higher grade fuel.

Heat and Combustion Considerations

The energy required to begin the combustion reaction comes from some initial source, such as a spark, or high compression of a fuel/air mixture. Provided that the other two elements, fuel and air/oxygen are present, the activation energy is supplied to begin the exothermic reactions. Exothermic refers to a net excess release of heat energy over that required to begin the combustion reaction.

The higher the temperature of the fuel/air mass, the more likely the combustion reaction will be maintained. It should be noted that the entire fuel and air mass is not required to be brought to the same temperature at the same time. Only a limited volume or mass of the fuel and air mixture needs to be brought to the ignition point, whereby combustion begins as a self-perpetuating wave front moving through the fuel and air mixture.

Left behind and exhausted are the products of combustion. Should the activation energy for complete combustion not be present, other products of incomplete combustion (PICs) may exist. The products of such incomplete combustion would include carbon monoxide, aldehydes, organic acids, and/or benzene fractions. Lack of sufficient oxygen is also responsible for products of incomplete combustion.

Air/Oxygen Considerations

Air typically supplies the oxidizer (oxygen) to the combustion process. Newer combustion systems have been proposed which provide pure oxygen (at a price), rather than utilizing air to provide the oxidizer for the combustion process.

When air is used in a system at very high temperatures, oxygen begins to react with the nitrogen in the air. This is a highly temperature-sensitive reaction, and the products become a significant contributor to the formation of photochemical air pollution. The overall reaction is:

$$^3/_2 \, O_2 + N_2 \rightarrow NO + NO_2$$

The amount of air per unit mass of fuel, how the air and fuel mix, and the operating temperature for a given combustion source largely determines the products of combustion; i.e., carbon monoxide, NO, aldehydes, etc.

COMBUSTION CHEMISTRY

The importance of combustion chemistry cannot be minimized in understanding the contribution of combustion products to the problem of air pollution. This is particularly true in areas with relatively high population densities.

The "mole" concept is a basic element of combustion chemistry and makes possible the computation of relative amounts of fuels, oxidizers, and combustion products. Equation 4.1 is an example of stoichiometric combustion of a given hydrocarbon fuel with air to yield (under ideal conditions) carbon dioxide, water, nitrogen (at about a 4:1 ratio with oxygen in air), plus excess heat. In the presence of excess air, there would be additional oxygen released in the exhaust products.

$$CH_4 + 2O_2 + 8N_2 \rightarrow CO_2 + 2\,H_2O + heat + 8\,N_2 \qquad (4.1)$$

This fuel-specific equation allows for computation of the theoretical mass of each combustion product based on the mole concept. However, it is possible, either on an overall basis or in certain portions of a combustion chamber, that a deficiency of oxygen would be present. In that event, an equation such as Equation 4.2 could be found:

$$CH_4 + \tfrac{1}{2} O_2 + 2 N_2 \rightarrow CO + H_2O + heat + 2N_2 + PICs \quad (4.2)$$

Such a starved air system yields CO and a number of partially oxidized fuel fragments as products of incomplete combustion (PICs) including aldehydes, acids, and higher molecular weight fuel fragments such as soot. Starved air situations are responsible for a number of the health effects associated with combustion emissions and photochemical ozone reactants.

In addition, small amounts of air will react as noted earlier to yield nitric oxide:

$$N_2 + O_2 + heat \rightarrow 2NO \quad (4.3)$$

The key to this reaction is that heat is required for this reaction to proceed, whereas the other reactions liberate heat.

Air to Fuel Ratio Considerations

One of the more important indices of combustion contaminant potential is the air to fuel ratio. This is a computational value of the ratio (by mass) of the theoretical amount of air required for complete combustion to a unit mass of a specific fuel. Air to fuel ratios are specific for each individual fuel. The perfect or *stoichiometric* amount of air for a given fuel mass is called the theoretical air. Typically, this ratio is in the range of 12 to 16 mass units of air per mass unit of liquid fossil fuels.

For a hydrocarbon fuel such as gasoline, this theoretical air to fuel ratio would be slightly above fifteen. Fuels with different percentages of carbon and hydrogen (and oxygen or inerts) would have somewhat different ratios. Air to fuel combustion ratios less than the theoretical amount are called *rich* conditions, since there is more fuel than needed. There would therefore be a deficiency of oxygen for complete combustion in such a system. Where there is an excess amount of air in a given combustion system, this would be considered a *lean* air to fuel ratio or an *excess air condition*.

If one is operating a combustion system with a rich mixture, i.e., a starved air system, one would expect a reaction similar to Equation 4.2 to predominate. Therefore, one would see higher values of carbon monoxide, products of incomplete combustion and other fuel fragments in the exhaust gases. As more air is added and the theoretically required amount approached, the concentrations of CO and PICs would diminish. If one kept increasing the amount of air for a given fuel flow rate, Equation 4.1 would begin to predominate. The combustion contaminants and fuel fragments would diminish to a minimal value. Operating in the lean regions of the air to fuel ratio, one would expect to see any fuel minerals and ash fully oxidized.

If the concentrations of exhaust gas contaminant concentrations are plotted against the air to fuel ratio one would see a curve similar to Figure 4.5. If one increased the air to fuel ratio beyond the stoichiometric ratio into the lean regime, the concentrations of CO and hydrocarbons begin to rise again. This is due to the temperature decreasing with increasing air, as air to fuel ratios approach about 18:1. Increasing amounts of fuel fragments and CO would not be burned and would be emitted.

The maximum temperature attained in a typical combustion system is at a peak slightly higher than the stoichiometric air to fuel ratio. As one moves away from the theoretical air to fuel ratio, the temperature attained in the combustion system drops off.

Oxides of nitrogen, also seen in Figure 4.5, tend to reach a peak just

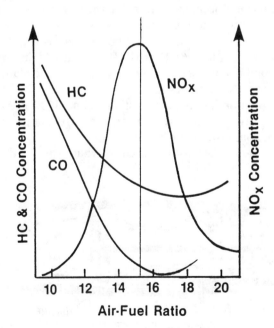

Figure 4.5.   The relationship of combustion products to I.C. engine A/F ratios.

slightly on the lean side of the theoretical air to fuel ratio, since that is the point at which the flame temperature is at its highest. In theoretical combustion experiments where no heat is lost from a combustor, this temperature can be measured and computed for ideal conditions. This is the *adiabatic flame temperature* and is the highest temperature that can be attained under stoichiometric air conditions. Equation 4.3 indicates the contribution of heat to the formation of oxides of nitrogen. When temperature is diminished in a combustion system there is less likelihood for Equation 4.3 to operate and therefore the oxides of nitrogen would be lower.

Likewise, if one were operating in the rich region, one would expect that fuel components would be subject to reducing reactions rather than being fully oxidized. Therefore, under those conditions, one would expect to see more carbon or partially oxidized fragments such as carbonyl sulfide, or even elements such as mercury (as a vapor).

Combustion Systems

Combustion systems generally fall into one of two categories: those that have the combustion occurring internal to the power system, and those that have the combustion external to the power system.

Typical of such an internal combustion system is a gas turbine engine (schematic seen in Figure 4.6). Here the flame is inside the combustor "can" and the hot combustion gases exhaust against a turbine blade. This drives a power shaft and also provides thrust.

The most common internal combustion engine (ICE) is the reciprocating internal combustion engine (RICE), and in particular, the spark ignited (S.I.) internal combustion engine.

The RICEs are further distinguished by what provides the initial activa-

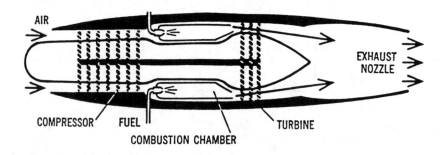

Figure 4.6.   Combustion in a gas turbine engine.

tion energy for the combustion system. In the typical gasoline powered automobile, this is provided by a spark plug. In a diesel system, high compression ratios drive the temperature and pressure of the mixture up until the fuel and air auto-ignite.

Internal combustion engines require cleaner fuels than external combustion systems; thus, they may fire gaseous fuels (propane or natural gas) or highly distilled liquid fuels such as kerosene or "jet-A". No minerals or ash components are permissible, since they must burn cleanly to allow the internal combustion engine to operate without fouling.

The external combustion system is the older system. In this combustion system, any fuel, solid, liquid, or gaseous, may be used for the heat needed to provide power to a working fluid. Therefore, a wide range of fuels with different ash contents, heating values, sulfur contents, etc., may be used with an external combustion system. Typical of such combustors would be boilers, heaters, fluidized beds, etc., in which the heat of combustion is transferred across a surface (usually pipes) to a "working fluid." Figure 4.7 shows a schematic of the most common external combustion system—a boiler. The working fluid, whether hot oil, water, or steam, is used to provide useful work in another location. Such a system could be for indirect heating, drying, melting, power production, or any combination of uses.

For regulatory purposes, there is a division of combustion equipment into stationary or mobile equipment. This distinction is more typically addressed under source control technologies. The emissions patterns are specific to the fuel, type of combustion, and air to fuel ratios, not whether the system is mobile or stationary.

EVAPORATIVE EMISSIONS

Evaporative emissions are those occurring from the loss of materials as a function of operating temperature and exposure to moving air. Therefore, the higher the vapor pressure and the exposure to ambient air, the greater the evaporation rate of such material. With the exception of elemental mercury emissions, organic solvents form the bulk of the evaporative emissions.

It is informative to review the effects of vapor pressure on an evaporative emission. Table 4.9 tabulates the vapor pressures of some common liquids at two different temperatures, 20° Celsius (°C) and 40°C. These are the most commonly used and evaporated materials in the United States. In all cases, as temperature increases, the vapor pressure of the respective organic material increases as well. It is not a direct function but a logarithmic one;

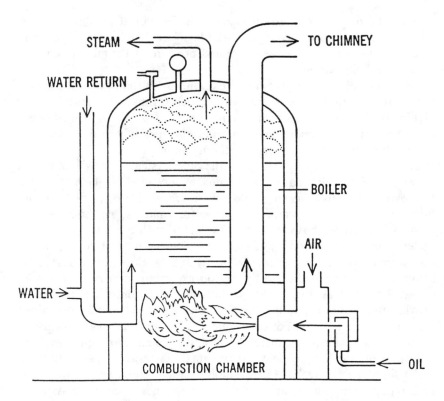

Figure 4.7.   A boiler schematic—the most common external combustion system.

Table 4.9. Vapor Pressure of Common Liquids (in mmHg)

Liquid	20°C	40°C
Acetone	185	424
Methanol	96	263
Methyl ethyl ketone	71	177
Toluene	21	58
Trichloroethane	106	257
Trichloroethylene	55	138
Xylene	5	17
Hexane	121	280
Octane	10	31

Adapted from *Handbook of Chemistry and Physics,* 72nd ed., CRC Press, Inc., Boca Raton, FL, 1992.

i.e., as the temperature increases, the vapor pressure increases at a faster rate. The temperature at which the vapor pressure equals 760 millimeters of mercury is the boiling point of the pure material at sea level.

At all temperatures, acetone has the highest vapor pressure of the common solvents in that table. Now, vapor pressure is just one measure of the tendency of the material to evaporate. The actual evaporation rate is a function of many variables including air velocity, exposed surface area, container color, surface area exposed to sunlight, etc. However, we can use vapor pressure as a first approximation of the tendency of various organic materials to evaporate.

Several of the common solvents, toluene and xylene, are also major components of gasoline. Therefore, it would be appropriate to review hexane (a C_6 compound), and octane (a C_8 compound), as well. Indeed, these four form a large portion of gasolines in general.

Components with higher vapor pressures at any given temperature tend to preferentially evaporate, leaving behind the lower vapor pressure materials. Thus, during handling of fuels such as gasoline, the lighter fractions of these four (hexane, with a vapor pressure of 121 mm Hg) would form a greater percentage in the vapors than in the total fuel composition. Likewise, the liquids left behind would show a deficit of these lighter, more easily volatilized components, and more become concentrated with xylene.

It is also interesting to note the differences between trichloroethane and trichloroethylene. At both temperatures trichloroethane has approximately twice the vapor pressure of trichloroethylene. Thus it is more volatile and will lose more material for a given situation than trichloroethylene.

Evaporative Classifications

The three major classifications of evaporative emissions by utilization are by:

- fuels production and processing
- coatings and solvents
- organic chemical production

Fuel production and distribution generate evaporative emissions of organic vapors due to the volatile nature of fractions of these materials. However, due to the monetary value associated with these organic materials as fuels, they tend to be more closely monitored and evaporative emissions minimized during production, processing, transport, and storage.

Coatings and the nature of coatings such as paints form a significant fraction of all evaporative emissions since the purpose of coatings is to deposit resins, pigments, and other agents over a surface, followed by loss of the carrier (solvent). The nature of a coating requires that the solvent be completely driven off before the coating can perform its primary role as a surface protection. Solvents act to dissolve, remove, or carry some other material. Cleaning of metal parts, functioning as a transport and spreading medium for paint or coatings, and carrying other materials, which are used as a chemical reactant, such as in the chemical processing industries, are basic functions of a solvent.

Organic chemical production involves synthesis of a wide variety of intermediate and final products from original raw materials. These processes may be either in a batch or continuous mode. Emissions occur from storage, production, and production of the materials.

There are a number of subdivisions possible for each of these major classifications, depending on the nature of the process or source. There are several hundred classifications of manufacturing operations that may contribute to evaporative emissions. These include petroleum refineries, fertilizer plants, aircraft parts and equipment, motor vehicles and truck body manufacturing, pulp mills, synthetic/organic chemicals, etc.

The federal Department of Labor has set up a classification scheme for all manufacturing facilities. These are known as the Standard Industrial Classifications (SICs) They are four-digit codes which refer to specific industries. In many regulatory schemes these SIC codes are utilized as source categories for regulation due to the similarity of the emissions from each process.

Fugitive Emissions

Fugitive emissions consist of primary pollutants generated by activities or processes other than combustion or evaporation. Thus a wide variety of emissions would be considered as fugitive emissions. Likewise, fugitive emissions encompass a number of different industrial and commercial activities. These include agricultural activities, food processing, mining, construction, steel making, rubber manufacturing, chemicals, stone, clay, glass, acid manufacturing, fertilizer manufacturing, and secondary metallurgical operations. Fugitive emissions include both gases and particulates.

The major activities generating fugitive particulate emissions are *materials handling* where primary pollutants are discharged. These include conveyors, storage devices, transfer points, and process equipment.

Processes and activities which give rise to *attrition* of the base materials also generate fugitive emissions. These include grinding operations, polishing operations, buffing operations, and sanding.

Those processes which entail a *fuming* process are also sources of fugitive emissions. Acid manufacturing plants are among those sources. Metallurgical operations in which hot metal is poured from a crucible into either a ladle or an ingot mold contribute to fugitive emissions. Other operations such as production of steel from a basic oxygen furnace (BOF) form fugitive emissions, even though "combustion" takes place in this process. Pure oxygen is injected into the molten pig iron in a BOF, at which point the carbon in the molten metal reacts to generate CO and CO_2 and particulates. These are blown out of the hot metal into open hoods and are conveyed by water-cooled ducts to control equipment.

Waste Related Emissions

From the perspective of the noncriteria pollutants, the issue of air emissions from waste related activities is an increasingly important one. Such activities include waste disposal facilities such as landfills or composting facilities, as well as activities involving site remediation, sewage treatment plants, and hazardous waste treatment, storage, and disposal facilities. The latter also includes incinerators.

Landfills

The interment of solid and hazardous waste materials in landfills has been an ongoing part of waste disposal practices for over 50 years in the United States. Associated with landfills are a wide variety of air contaminants, which pose a potential health threat to residents in the immediate vicinity of such land disposal facilities. The problem relates to the fact that not only are there small quantities of toxic or hazardous materials in common household trash, but for a number of decades, hazardous waste from industrial sources was co-disposed with municipal waste. As a result, landfills tend to have a unique mix of air contaminant emissions. The majority of these relate to hazardous air pollutants.

There are three major categories of air pollutants being emitted from landfills: *biologically* generated, *evaporative*, and those that are a result of *smoldering combustion* within the landfill mass itself.

The biologically generated air pollutants include those which are off gases from metabolic activity of bacteria living in anaerobic conditions in a landfill and are "processing" decomposable wastes. These gases include methane, carbon monoxide, and metabolically altered contaminants such as vinyl chloride (from the action of bacteria on perchloro- and trichloro-ethylene).

Early in the evolution of emissions from an active landfill will be emissions of carbon monoxide and hydrogen gas. After the first several years of operation, a typical landfill will be generating landfill gas with a composition of approximately 50% methane and 50% carbon dioxide. However, trace quantities of air toxics will also be a part of that landfill gas (in the ppm range). Reduced sulfur compounds can also be expected from the metabolism of sulfur-containing materials by bacteria to yield compounds such as hydrogen sulfide. This results from bacteria acting on materials such as calcium sulfate in gypsum or wallboard, which is typical of construction and demolition debris.

Figure 4.8 is a schematic of a typical landfill showing the emissions from the surface and subsurface on their migration pathways. Landfills form an area-wide source of air contaminants. Figure 4.9 is a schematic showing the generation curves for landfill gas over time. Ultimate yields of methane are estimated to be between 1 and 4 cubic feet per pound of municipal waste. It should be remembered that each of the localized areas of solid and hazardous waste disposal will have begun their emission curve history at different points in time. Therefore, samples of landfill gas from different points in the landfill will consist of varying gas constituents.

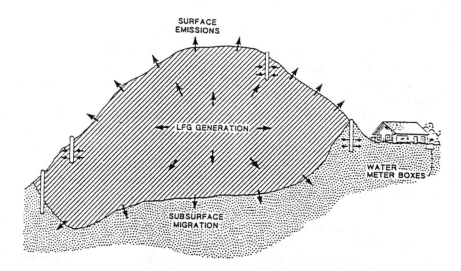

Figure 4.8.   Schematic of landfill with LFG movement.

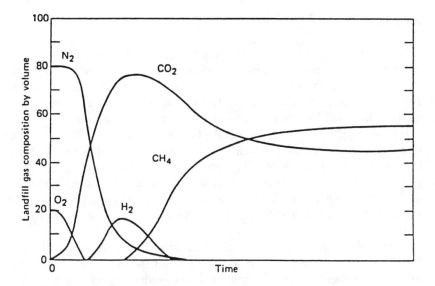

Figure 4.9.   Landfill gas composition versus time.

Also, some part of the landfill gas will be formed of vaporized solvents or liquids which have been co-disposed with the hazardous waste fractions. Evaporative emissions will be those of the higher vapor pressure, lower molecular weight organic materials such as acetone and methyl ethyl ketone, or toxics such as benzene. Actual emissions will vary widely, depending upon the amounts disposed in each individual landfill.

A third mechanism for air contaminant emissions from landfills would be those from *smoldering combustion*. In this situation, smoldering fires exist within the landfill mass generated from spontaneous combustion of combustible gases and infiltrating air. It is not uncommon to find core temperatures in the middle of landfills as high as 175°F.

Air is introduced into a landfill mass by subsidence and/or surface cracks where it encounters methane. "Starved air" combustion begins. The gases driven off will be a variety of partially oxidized and reduced components of fairly low molecular weight. Such contaminants include carbon monoxide, hydrogen sulfide, carbonyl sulfide, and other products of incomplete combustion. These emissions tend to be localized and will vary widely in composition and mass emission rate from one portion of a landfill to another. Emissions at different landfill sites within the same region will be significantly different.

Treatment Related Emissions

Among the increasing sources of waste related air contaminant emissions are those operations which involve some aspect of hazardous waste site remediation. This may range from a major Superfund site to a small vadose zone cleanup of a gasoline spill from a local filling station. Therefore, these air emissions consist more of toxics than of criteria pollutants.

Handling of soil, whether by well drilling, soil removal, or movement for treatment, generates dust which could be contaminated with the entire range of hazardous chemicals. This could include benzene, polychlorinated biphenyls, lead, etc. Groundwater treatment, whether in situ or pumped and treated, may also be a source of volatile organic compounds as from vapor extraction, unless the soil gases are vented to an air pollution control system.

Waste treatment, storage, and disposal facilities are likewise sources of air contaminant emissions. Due to the nature of the operation, these emissions cover the entire range of combustion, evaporative, and fugitive emissions of both criteria and noncriteria contaminants.

Landfills can be a source of particulates, as well as evaporative and fugitive emissions. Incinerators present the full range of air pollutant emissions from a combustion process, and have the potential for evaporative emissions of the feed materials containing hazardous contaminants. Evaporative emissions from treatment facilities include solvents or reduced carbon gases such as from leachate or surface ground treatment facilities. Additives from storage facilities and transfer points may generate fine dusts.

Municipal water treatment facilities may also be a source of air toxic emissions resulting from evaporation of low molecular weight organics and odors from the water being treated. The source is from solvents and paints dumped into the sewer. The wastewater carries the materials to the sewer treatment plant. During the handling of water for primary, secondary, or tertiary treatment, these emissions are volatilized and lost to the atmosphere. Control of these emissions, in particular odor-causing contaminants commonly associated with sewer plants, is a large part of the local concern for emissions from these facilities.

CRITERIA AIR POLLUTANT GENERATION

The anthropogenic criteria air pollutants are formed by a variety of processes. These correspond generally to their major classifications.

Oxides of nitrogen (NO_x) are overwhelmingly formed by combustion processes involving air and high temperatures. The principal contaminant is nitric oxide (NO), usually with smaller percentages of the total being nitrogen dioxide (NO_2). In certain liquid and solid fuels where there is fuel-bound nitrogen, up to 50% of that fuel-bound nitrogen may be emitted as NO_x. For a few localized industrial sources, there are discharges of HNO_3. These include acid manufacturing plants, plating shops, and chemical operations where the reactant is NO_2.

Sulfur dioxide is formed predominantly by oxidation of fuel or product materials containing sulfur. Sulfur in liquid or solid fuels comes from the reduced sulfur compounds inherent in the fossil fuel. In coal, pyrites (iron sulfide) are intimately mixed with the soils interbedded with coal seams. Product sulfur is typically associated with the chemical process industry where sulfur-containing chemicals are oxidized and/or directly emitted. Refineries emit sulfur oxides from the cracking and reforming processes used to produce gasoline.

Carbon monoxide is formed by *quenched combustion*. This includes situations where burners are clogged, eroded, or poorly maintained. CO also forms in combustors where localized deficiencies of oxygen occur, as on solid fuel grates. Operation at a fuel-rich combustion ratio contributes to CO formation. There are a few fugitive sources such as "young" landfills, which produce carbon monoxide in the early stages of waste decomposition.

Hydrocarbons or organic gases are emitted either as evaporative emissions or products of incomplete combustion. Other minor and localized sources include those hydrocarbons which could escape from an air pollution control device by virtue of a slipstream or malfunctioning control system.

Particulate emissions may be either fugitive or formed by combustion. The latter include combustion contaminants such as soot or ash. Fugitive emissions occur from size classification operations or mechanical handling of solid materials like coal, limestone, or cement. Processes generating fumes such as sulfuric acid production or metallurgical operations during pouring of hot metal also generate particulate matter.

Lead is emitted primarily by industrial processes or as a result of leaded gasoline use. It is therefore both a direct emission and a result of fuel additives in combustion. Lead, being an element, is neither created nor destroyed, but may change oxidation states such as from liquid (tetraethyl lead) to a solid particulate such as lead oxide.

Ozone, as a secondary pollutant, is formed photochemically in the atmosphere from the reactions of NO_2 and organic gases in the presence of sunlight.

Source Inventories of Criteria Pollutants

Air quality management strategies require that we understand the basic source classifications of air contaminants and the emission rates of criteria pollutants. These *regulatory classifications* may be made in a number of different ways: by pollutant, by process, by SIC code, by number of sources, or by emission rate, among others.

The U.S. EPA has adopted a five category scheme which classifies the nation's emissions into five major categories. These are:

- transportation
- fuels combustion
- industrial processes
- solid waste, and
- miscellaneous

Table 4.10 lists the 1991 man-made criteria air emissions inventory estimates by category for each of the criteria air pollutants. Each of these classifications forms a convenient approach to summarizing the nation's emissions.

Transportation

Transportation includes combustion, evaporative, and fugitive emissions from mobile sources. The types of sources include aircraft, railroad, ship emissions (within harbor areas and rivers), those from trucks or heavy-duty vehicles, and light duty gasoline powered automobiles. All of the anthropogenic emissions are represented in the transportation category.

Table 4.10. 1991 U.S. Emissions by Source Classification[a]

Source	CO	PM10	NO	SO_2	VOC	Lead
Transportation	43.5	1.51	7.3	1	5.1	1.62
Combustion	4.7	1.10	10.6	6.5	0.7	0.45
Industrial	4.7	1.84	0.6	3.2	7.8	2.21
Solid Waste	2.1	0.26	0.1	0	0.7	0.69
Miscellaneous	7.2	0.73	0.2	0	2.6	0
Totals:	62.2	5.45	18.8	20.7	16.9	4.97

[a]In millions of tons per year, except lead (thousands).
Source: Natural Air Quality and Emissions Trends Report, 1991. USEPA Office of Air Quality Planning and Standards, 450–R–92–001, October 1992.

For the purposes of management and regulation, transportation sources are considered separately from stationary sources, since mobile sources cross state boundaries.

Transportation represents the largest single category of CO emissions (2/3 of the nation's total), nearly 40% of the NO_x, 30% of the hydrocarbons, 28% of the particulates, and 32% of the lead. These are due primarily to internal combustion power systems.

Combustion

The fuel combustion classification predominates among the stationary sources of air contaminant emissions, due to the tremendous amount of energy utilized in the United States. These stationary sources of fuel combustion can be divided into two groups; those which produce electrical power, and those which use the heat of combustion in some other fashion. The former includes coal, oil, and gas-fired utility boilers providing electrical power to the nation. The latter include all other stationary sources of fuel combustion such as steam boilers, heater treaters, and those operations which require process heat. These include melting or smelting operations (metallurgical activity), those which provide heat for curing, drying, or firing (cement kilns), or those which provide high temperature to a heat transfer fluid such as in the chemical process industries.

Fuel combustion is the primary source of sulfur oxide emissions and the major source of oxides of nitrogen. Combustion contaminants contribute significantly to PM10 emissions. Carbon monoxide levels are fairly low in the inventory since CO represents an economic loss due to poor fuel efficiency. Therefore, stationary sources of fuel combustion tend to tightly control CO and hydrocarbon emissions by combustion modifications. Lead emissions are low. This is due to the phaseout of lead additives in all fuels including gasoline.

Industrial

The third source classification is industrial processes whose emissions are of a noncombustion nature. A schematic of a relatively simple operation for an industrial source, such as glass manufacturing, is shown in Figure 4.10. In this figure we see a number of source emissions in a relatively simple process. Typical of these emissions would be those lost by fugitive sources such as handling of materials from storage and combustion emissions from a furnace used to melt the glass ingredients.

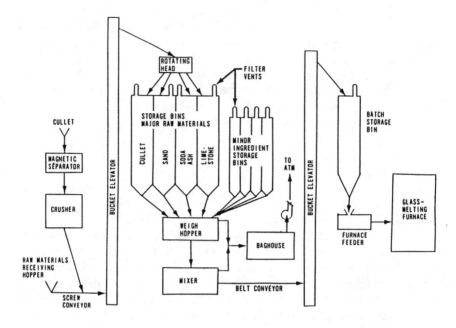

Figure 4.10.   Process flow diagram of a batch plant.

The major contaminant from industrial emissions is volatile organic compounds. These include all operations involving solvents, cleaning and coatings of materials, or those processes which manufacture chemicals, solvents, cleaning agents, or surface coatings.

Industrial processes are significant PM10 generators due to the large number of materials handling operations. Industrial sources are second in terms of total emissions of sulfur dioxide. These are due primarily to industrial processes which release sulfur oxides from a noncombustion source, i.e., sulfuric acid plants, ore roasting and mining, or smelting operations.

Waste and Related

The solid waste and related classification includes waste disposal, hazardous waste incineration, and landfill activities. Overall, this is a fairly small category due to the relatively small number of sources.

Miscellaneous

This classification includes estimates for miscellaneous emissions from consumer products usage, home usage of volatile materials from paints, cleaning materials, cigarette usage, fireplace combustion, etc.

Comparisons by Category

In Figure 4.11, the source-generating categories are shown by contaminant versus emissions in millions of tons per year. Lead is not shown because the total emissions are less than 5,000 tons per year. The majority of the CO comes from transportation. At 62 million tons per year, CO is the largest single contaminant emitted by mass in the United States.

Table 4.11 indicates the different source categories and emissions for natural fugitive emissions of PM10. These are due to the action of weather on existing materials. It should be noted that these totals are over eight times the anthropogenic emissions attributed to the five major source categories in Table 4.10. This may have significant impacts on the air quality management strategies adopted in the future for PM10 concentrations.

Of all of the pollutant categories, the most significant changes have been in the emissions of lead, with approximately 90% of the 1982 levels being reduced to the 1991 levels of less than 5,000 tons per year.

HAZARDOUS AIR EMISSIONS

Of recent concern have been the emissions of hazardous air pollutants. These are not criteria contaminants, but do contribute to increased risks of adverse health impacts with potentially fatal results. These include toxics as well as carcinogens, mutagens, and teratogens.

The major source of our information on air toxics comes from the 1986 Emergency Planning and Community Right to Know Act (EPCRA). Under Section 313, it requires that hazardous materials emissions to all media from certain industrial sources be quantified. These emissions have been quantified annually since 1987 for the Section 313 chemicals. Table 4.12, from the 1991 report, shows the top 10 hazardous air emissions released and reported that year to the EPA.

It is interesting to note that the EPCRA values are reported in millions of *pounds* in the annual Toxics Release Inventory, not millions of tons. HAPs

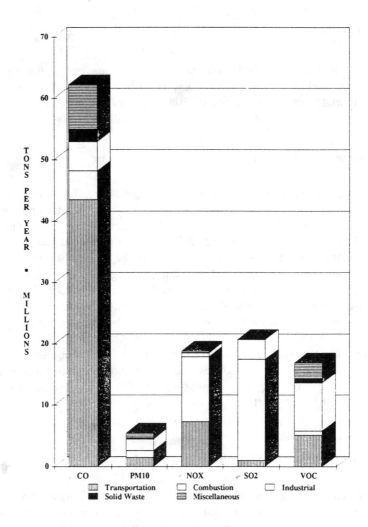

Figure 4.11.   EPA source inventory categories (by pollutant).

are released at a much lower rate than criteria pollutants. Indeed, approximately three orders of magnitude lower emissions are attributed to HAPs.

Also, these are releases from only the 20 SIC coded industrial classifications required by EPCRA, and do not include the emissions from mobile sources such as aircraft, railroads, and automobiles, or other industrial sources. With the exception of hydrochloric acid, all of these compounds are volatile organic gases. Also, it should be noted that the emissions of

Table 4.11. Natural PM-10 Fugitive Emission Estimates, 1986–1991

Source Category	Million Metric Tons/Year					
	1986	1987	1988	1989	1990	1991
Agricultural tilling	6.26	6.36	6.43	6.29	6.35	6.32
Construction	10.73	11.00	10.58	10.22	9.11	8.77
Mining and quarrying	0.28	0.34	0.31	0.35	0.34	0.36
Paved roads	6.18	6.47	6.91	6.72	6.83	7.39
Unpaved roads	13.30	12.65	14.17	13.91	14.20	14.36
Wind erosion	8.52	1.32	15.88	10.73	3.80	9.19
Total	45.27	38.14	54.28	48.22	40.63	46.38

Note: The sums of subcategories may not equal total, due to rounding.

each of these compounds have been reduced significantly since the original base year of 1987, without legislative or regulative activity.

QUANTIFICATION OF SOURCE EMISSIONS

Source emissions may be quantified in a variety of ways, each with various levels of accuracy and related costs. Cost implications of quantifying HAPs become a concern due to the high degree of specificity, precision, and accuracy required by the low emission rates and exceedingly low concentrations normally associated with air toxics. Those contaminants which are

Table 4.12. Top 10 Hazardous Chemical Air Releases,[a] 1991
(tons per year, 1000s)

Chemical	Fugitive	Stack	Total
Methanol	17.9	81.9	99.8
Toluene	36.8	62.5	99.3
Ammonia	23.5	70.8	94.3
Acetone	42.3	37.8	80.1
1,1,1-Trichloroethane	34.6	34.2	68.8
Xylenes	13.8	44.0	57.8
Methyl ethyl ketone	16.5	35.2	51.7
Carbon disulfide	1.3	43.4	44.7
Hydrochloric acid	2.3	39.2	41.4
Dichloromethane	15.9	23.8	39.7

[a]USEPA, 1991 Toxics Release Inventory, Public Data Release.

solid in nature can be analyzed from the total particulate emissions collected in the process or stack gas. Each quantification technique may or may not be appropriate for a given source.

Source Testing

The first of the quantification techniques is termed source or emission testing. This method is appropriate for all categories of air contaminant emissions except for fugitive sources. Therefore, combustion emissions, evaporative emissions, and direct emissions may be tested using source testing methods, provided that the emission is contained in a stack, chimney, or exhaust duct. Source testing methods are published by the U.S. Environmental Protection Agency.

Source testing requires measurements of all of the fluid flow characteristics of the exhaust gas. This includes temperature, pressure, moisture, and velocity of the gas flow at different diameters of the duct or stack. Appropriate methods of collection for the contaminants in that exhaust stream are also to be followed. Each of the test methods is specific to a source, contaminant, or a class of contaminants such as particulates or total hydrocarbons, and type of source.

When the contaminants of interest are pure gases, the test methods allow for measurements of gas flow followed by sample collection as independent efforts. When the contaminants are particulates, the sample collection must be at the same velocity as the exhaust gases. This is termed *isokinetic sampling*.

The standardized methods are required by the EPA for determining compliance with source emissions standards. Other test methods may be used in specific cases or at the request of state or regulatory officials. At best, the accuracy of this method is $\pm 15\%$ to 20% of the source emission strength. Poor technique, collection efficiency, lab handling, and analysis techniques may substantially impact the accuracy and precision of a stack test.

Carbon Balance

The *carbon mass balance* approach is appropriate only to combustion sources for simple gaseous emissions. This technique allows measurement of the total gas flow by utilizing accurate fuel flow measurements, combined with a precision analysis of the fuel being used. It assumes that air is the oxidizing medium. The technique depends on a stoichiometric mass carbon balance for its accuracy.

Table 4.13. Source Emission Quantification Approaches

Approach	Accuracy[a]	Combustion	Evaporation	Direct	Fugitives
Source testing	15–20	XXX	XXX	XX	X
Carbon balance	3–5	XXX			
Composition	5–15		XXX		X
Emission factors	50–150	X	X	X	X

XX, etc. = Relative degree of applicability.
[a]Relative percent accuracy, plus or minus.

One measures the contaminant concentrations and the excess oxygen levels in the stack at the same time the fuel flow is being monitored. Fuel flow, excess air and fuel analysis allows the calculation of total gas flow. Gas flow times the pollutant concentration gives the mass emission rate. Accuracy levels are limited only by the accuracy of the fuel flow meter, excess air measurements in the stack, and the fuel analysis. Typical accuracy ranges are ±3% to 5%.

Composition

The most common technique is calculation based upon *composition*. These are primarily directed at evaporative sources where one knows the compositional analysis of the raw materials, the total usage of material, and the percent evaporated or discharged. The levels here are ±5% to 15% accuracy, with much lower accuracy when applied to fugitive sources. The fugitives are typically particulates.

Emission Factors

The fourth major technique is actually a combination of all of the previous ones. This technique combines them all into an average *emission factor* for a given process. Therefore, it relies upon tests of similar equipment (which vary widely in emission characteristics). The ultimate result is a process-specific factor, typically contaminant mass emissions per unit output. This is multiplied by the total throughput or production rate to yield the total mass emission rate.

This technique allows calculation of mass emissions only for a *category* of sources, and therefore has a much lower level of accuracy. Typically, these factors are ±50% to 100% for a given source. Therefore, it can only be used for estimates on a large number of sources over a geographic area, or an annual period for emission inventories and air quality management strategy development.

Fugitives

Fugitive emissions typically are based upon calculation, or in a few cases, by field testing of downwind concentrations during appropriate wind conditions. Such techniques leave large concerns for accuracy of emissions from sources such as coal piles, landfills, and storage of dusty materials.

Table 4.13 is a summary of the accuracy of each source category appropriate to each of these quantification techniques.

5 METEOROLOGY, DISPERSION, AND MODELING

Who has seen the wind? Neither you nor I: But when the trees bow down their heads, the Wind is passing by.
Christina Rossetti, *Who Has Seen the Wind?*

The transport and dispersal of air contaminants are strong functions of the wind. To better understand such transport, dispersal, and dilution it is necessary to gain an understanding of the movement of the atmosphere on a local, regional, and global scale.

The thin band of our atmosphere makes it possible for life to exist on the face of the earth. The source of the energy which provides livable conditions is the sun. The absorption of that energy is responsible for the temperature differences noted on the earth's surface. These differences are responsible for the movement of the wind. Wind movement is complicated by the structure of the earth; i.e., the pattern of oceans, mountains, and continents at various latitudes.

The earth's energy exchange and the movement of the wind as they relate to air pollution are the subject of this section. The structure of the atmosphere, its dispersion characteristics, and the interplay of point, line, and area sources of air contaminants are significant components in our understanding of pollutant dispersion. From this understanding we are able to model the movement of air parcels and the dispersal of primary air contaminants. Modeling the results of photochemical reactions simultaneously occurring in the atmosphere is the logical next step in attempting to proactively manage our air quality resources.

113

EARTH'S ENERGY AND AIR MOVEMENT

The sun is the source of all energy received on the earth and it is instructive to evaluate the spectral distribution of that radiation—both as it is received at the top of the atmosphere, and as it is seen at surface level. Figure 5.1 is a plot of solar radiation intensity versus wave length for the two levels. The difference between the top and bottom curves is the amount of energy absorbed by different components of the atmosphere. The majority of this absorption is due to oxygen, ozone, water vapor, and carbon dioxide. Incident radiation varies as a function of the sun angle and latitude.

A net radiation diagram is seen in Figure 5.2 which indicates the mean irradiation of the earth at various latitudes in calories per square centimeter per day. The maximum occurs at the equator and the minimum at the poles.

Of the total amount of solar energy entering the earth's atmosphere, only about 53% is available at the ground level after scattering, absorption, and reflection processes. The ground exchanges energy by radiation, by evaporation and condensation of water, by exchange of sensible heat between the surface and the air, and by conduction into and out of the soil.

During the day, the surface has a net influx of radiation. During the night it loses a net quantity of radiation. During the day, when the ground is warmer than the air, heat is transferred from the ground to the air by convection. At night, the air is usually warmer than the ground so the convective transfer of heat is from air to ground.

Evaporation of water away from the surface requires both a moisture gradient (or concentration difference) as well as sufficient energy to supply the heat of vaporization necessary for water to change its state from liquid to gas.

If the ground is very dry, the net radiational input during the day will go primarily into convection and conduction. The atmosphere will be turbulent and windy, as is seen in deserts. If the ground is moist or the vegetation well watered, evapotranspiration will consume the major fraction of net radiation and the atmosphere will be quiet. These are basically physical processes related to the thermodynamics of the earth's surface and to conditions of climate.

Temperature and Global Air Movements

More than just molecular absorption by the air is responsible for the reduction in incident solar energy received at the earth's surface. These other

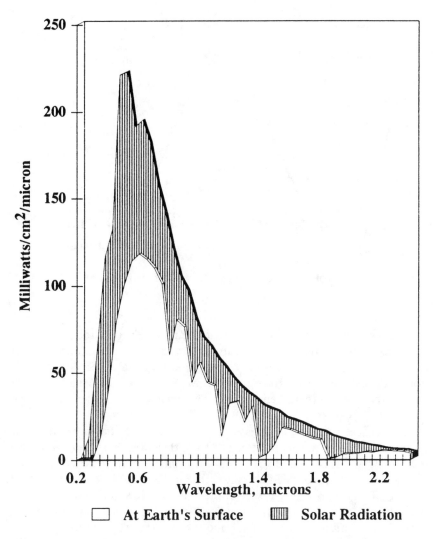

Figure 5.1.   Spectral distribution of solar radiation reaching the earth at the top and bottom of the atmosphere.

processes include reflection from clouds, diffuse scattering, and absorption by particles, primarily particulate aerosols such as sulfates in the upper atmosphere. Clouds are important from a global perspective, since at any point in time they cover approximately two-thirds of the surface of the globe.

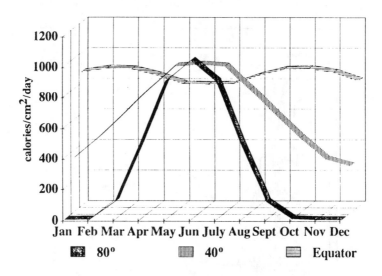

Figure 5.2.   Seasonal variation of insolation versus latitude.

As a result of the differential energies being received at the earth's surface, there is a difference in temperatures. On the local level, through direct contact with the earth's surface, or through absorption of terrestrial radiation, the atmosphere is warmed from below. When air is hotter in the lower regions than above, it will tend to rise, which will cause a general *overturning* of the parcels of air.

This vertical convection of gases gives the name to that portion of the atmosphere, the troposphere. It comes from the Greek word *tropos*, meaning "to turn." This local scale effect may carry up to a dozen or more kilometers in altitude, which we may consider the lower troposphere. These are not diffusion processes as seen at the molecular level, but rather convective motions extending over many kilometers. This action superimposes a general circulation pattern on air movements over the entire globe. It turns out there are sustained average patterns of movement of air in the troposphere.

Global Circulation Cells

In 1735 Sir George Hadley first proposed a model illustrating the above effects. Such a model is seen in Figure 5.3, in which the air, hottest at the

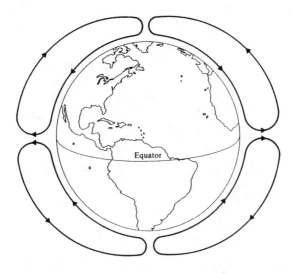

Equator

Figure 5.3.   Hadley's general circulation scheme.

equator, moves toward the poles, while the colder air at the surface of the poles tends to sink (being denser). It therefore moves toward the equator.

Modern observations of the actual movements of the wind at altitude as well as surface vectors reveal a more subtle and complex picture. The vertical and horizontal motions approximated in the Hadley Model occur from the equator to slightly above 30 degrees latitude, and again at latitudes above about 60 degrees. However, there appears to be another counter-flow cell of a weaker nature between the two. This is the Ferrel or mid-latitude cell. This mid-latitudinal cell is not as well developed.

In Figure 5.4 we see a cross-section showing these three cells. The three zone model, an adaptation of the Hadley model, attempts to explain the different air movements actually measured. Areas of typically low wind speed are called the equatorial doldrums, the horse latitudes (at about 30 degrees), and the subpolar lows. The polar easterlies, the westerlies across the temperate zones, and the trade winds (between the equator and approximately 30 degrees latitude) are also accounted for in this model. Lower level air near the equator has a high humidity so when it rises, cools, and condenses, clouds are formed with resulting precipitation.

To either side of the equator is the fairly well established Hadley, or trade-wind cell, in which tropical air rises as a result of its absorption of

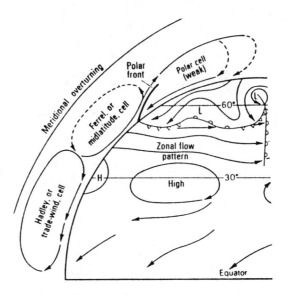

Figure 5.4.   Three-cell wind system.

heat and resulting gain in buoyancy. As warm air moves toward the poles at high altitudes, it loses heat by thermal radiation. This decrease in temperature of several degrees Fahrenheit per day causes a loss in buoyancy. At latitudes of approximately 30 to 32 degrees, a generalized *subsidence* of the air occurs in what are termed the subtropics. This global subsidence has important consequences for general circulation, as well as presenting a strong *summertime potential for air pollution* in the west coastal communities of continents. Once this subsiding air mass reaches lower tropospheric levels the air takes either an equatorial path to complete the Hadley cell or continues on a poleward route. This zone at or about 30 to 32 degrees latitude causes the doldrums in which the winds are gentle or nonexistent for significant periods. There is little circulation across such doldrums.

Above approximately 60 degrees latitude, a circulation zone called the polar easterlies occurs in which warmed air from the mid-latitudes moves northward at high altitudes, cools, and begins subsiding again over the polar regions. The circulation at this zone is completed at lower altitudes as air moves toward the temperate zone to take the place of the rising air. The weather patterns of the tropical and polar regions are, therefore, fairly well established.

By contrast, the temperate regions appear to have the most variable weather patterns. These are belts extending between about 35 and 55 degrees latitude where the polar and subtropical influences interact. These circulation patterns are not as well defined. In this region, energy is transported through the temperate zone by large-scale turbulence.

Jet Streams

In the temperate zones at the interfaces of the Ferrel cell with both the Hadley and the polar cells, there are discontinuities in the tropopause (Figure 5.5). These discontinuities are the locations of high velocity "rivers of air" called the *jet streams*. There is both a subtropical and a polar jet stream. It appears that through these discontinuities much of the circulation occurs between the stratosphere and the troposphere. Thus, much of the material injected into the stratosphere during violent volcanic eruptions works its way through this discontinuity before it can be brought to earth through convective motions within the troposphere. Likewise, this is the point at which stratospheric ozone appears to be injected into the troposphere. The polar jet stream, being less well developed, tends to meander considerably more than the subtropical jet stream. It is also influenced considerably by the effects of continental land masses.

Surface Effects

Even with these as simplistic overviews, there are significant variations in general air movement due to the structure of the earth's surface. These are due primarily to the differences in land and sea masses between the two hemispheres. The Northern Hemisphere contains the preponderance of the land mass, whereas the Southern Hemisphere contains the preponderance of ocean. Thus, the wind flow patterns are significantly different. The irregular patterns in the Northern Hemisphere are much more apparent than seen in the Southern Hemisphere. This is due to the vertical obstructions represented by continental mountain ranges and land masses, and the highly variable surface temperatures between ocean and land.

The poles themselves vary considerably in their structures as well. The north pole is an ocean (albeit covered with ice) surrounded by land masses, whereas the south pole is a continental land mass surrounded by open oceans. Other significant differences are that Antarctica has an average height between 7,000 and 8,000 feet, whereas the north pole is less than 1,000 feet. Additionally, the average winter temperatures of the north polar region are approximately -31°F, whereas over the Antarctic continent, tem-

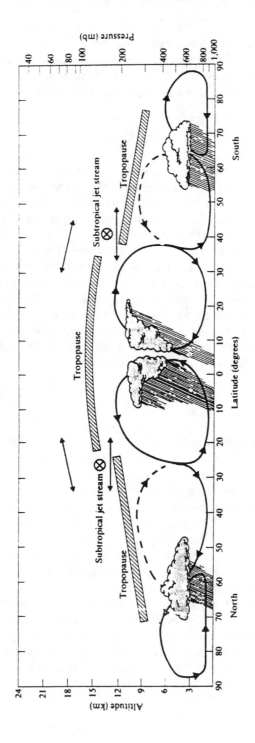

Figure 5.5.   Northern Hemisphere summer circulatory system in the troposphere.

peratures range from -40° to -94°F. Therefore, the air mass over Antarctica is considerably thinner, colder, and drier than the mass of air over the north pole. These factors may have significant implications for global air pollution and stratospheric ozone.

Other Forces

In addition to these general wind movements, there are other forces acting on the atmosphere on a global scale.

One force, called the Coriolis force, or effect, named in honor of Gaspar de Coriolis (1835), shows the additional impact of the earth's rotation on the macro scale patterns of air currents. Due to the earth's rotation, an apparent turning of the air in a rotational manner occurs due to inertia. The equator of the earth is moving at approximately 1,000 miles per hour, whereas the higher latitudes are moving at somewhat less velocity. Consequently, this rotational force is an inertial influence, acting in both northern and southern hemispheres.

In the northern hemisphere, the Coriolis force imparts a clockwise rotation to any moving air mass. In the southern hemisphere, the Coriolis force imparts a counterclockwise motion to the atmosphere. The Coriolis effect is a relatively weak force so its effects are only observed in large-scale wind patterns where, over a long period of time, the small acceleration due to its effects can produce an appreciable change in wind velocity.

Frictional forces between the moving air and the earth's surface must be added to the Coriolis effect to predict the response of moving air masses.

Patterns of High and Low Pressure

Standing patterns of high and low pressure regions develop over the earth's surface as a net result of the above effects. These are statistical averages over long periods of time, and therefore, can only be considered as an average condition.

A lower air pressure region along the equator corresponds to the zone of tropical air at the juncture of the Hadley cells. Moving away from the equator are zones of high pressure. These subtropical high pressure units are centered over the oceans in each hemisphere, primarily due to the influence of continental land masses. The most prominent is the eastern Pacific or Hawaiian high pressure zone. These higher pressure zones tend to show a clockwise rotation in the northern hemisphere and a counterclockwise rotation in the southern hemisphere.

Winter is the period when each hemisphere is experiencing its lowest sun angle. In winter conditions higher latitude low pressure regions push closer to the equator. This is due to the inclination of the earth's axis of rotation superimposed on the three zone-Hadley cell model mentioned earlier. The influence of these different continental pressure regions is responsible for much of the weather and air pollution potential experienced on the land masses.

In the temperate regions, the summertime global belts of subtropical high pressure are separated into individual regions centered over the oceans. Continental low pressure regions form over the southwestern United States, south central Asia, and Australia during their respective summer seasons. In these areas, the arid regions are warmed by solar heating sufficiently to produce strong vertical thermal currents which rise to high altitudes and impede the subsidence of air in the middle or upper troposphere. The corresponding column of rising warm continental air is less dense than the surrounding air over the oceans, so that a lower pressure region develops. These lower pressure regions are termed *cyclones*, due to the spiraling or vortex nature of the wind flows. Depicted in Figure 5.6 is the horizontal and vertical cross-section of such a low pressure zone (or a cyclone) in the northern hemisphere. They are normally accompanied by cloudy skies, precipitation, and considerable turbulence, which enhances the dispersion of air contaminants.

High pressure regions are termed *anticyclones*, since they show air movements in the opposite direction. Anticyclones are produced by regions of higher pressure, where cool air descends from aloft and diverges outward in

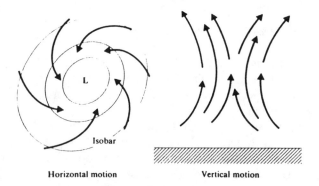

Horizontal motion   Vertical motion

Figure 5.6.   Low-level counterclockwise spiral of winds which converge in a cyclone in the Northern Hemisphere. The vertical motion of the air is depicted at the right.

a spiraling manner. Figure 5.7 is a cross-section of the horizontal and vertical pictures for such a high pressure system.

Anticyclones significantly affect the dispersion of pollutants over large
regions. As the air in a high pressure system descends, it is warmed by
compression. In the lower regions of a high pressure system, the air has a
higher temperature than the cooler air parcels directly in contact with the
earth's surface. This results in a *subsidence inversion*.

In the northern hemisphere, on the easterly side of the Hawaiian or
Eastern Pacific high, the inversion layer dips closer to the surface with
increasing distance away from the cell's center. As a result, the west coast of
north America experiences relatively low subsidence inversions. Consequently, areas such as southern California will experience an inversion base
at less than 2,000 feet for the majority of the summer months. Such subsidence inversions significantly reduce vertical air movement and pollutant
dispersion.

Due to the differences in pressure at various locations, we find that the
air flow tends to move laterally from areas of high pressure to lower pressure. This is called the pressure gradient, or the pressure gradient force.
Where pressure gradient forces are large, high surface winds may result.
Where the pressure gradient is small, surface winds are light.

In general, cyclones, or low pressure areas tend to move rapidly, whereas
high pressure zones or anticyclones tend to have a semipermanent nature.

High pressure anticyclones over land consist of subsiding air in the upper
regions, and are generally accompanied by clear skies and fair weather. Due
to the lower average speeds of anticyclones, it may also be possible for the

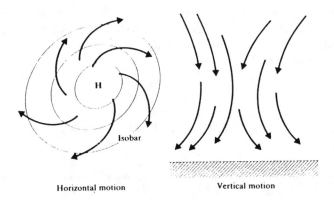

Horizontal motion   Vertical motion

**Figure 5.7.   Clockwise diverging spiral of winds from an anticyclone in the
Northern Hemisphere. The vertically subsiding motion of the
air is shown at the right.**

high pressure to practically stop their movement at certain times of the year and "stagnate." These may be either warm or cold anticyclones.

When anticyclones stagnate, we find classical air pollution episodes occurring. In stagnation conditions high pressure regions are nearly stationary and winds are exceptionally light. Such stagnations are common in the southeastern mountain section of the United States, as well as the Great Basin valley in the West, and the San Joaquin Valley in California. Figure 5.8 indicates the total number of air stagnation days in the eastern U.S. during a 30-year period when these stagnation conditions occurred in groups of four or more days. These conditions are a contributing factor to the frequent accumulation of reactive hydrocarbons given off by trees and vegetation in those areas. The resulting haze is considered responsible for the Great Smoky Mountains in eastern Tennessee, and the Blue Ridge Mountains over the Virginias and North Carolina.

Friction

For an accurate description of the motion of air at low levels, the friction of air with the earth cannot be neglected. Air immediately next to the ground is hindered in its motion by surface irregularities, which give rise to complex mechanisms of air movement. Friction gives rise to turbulent motion, which causes a transfer of momentum between the earth and the air. Where high pressure regions exist, the effect of friction causes the air to move from regions of higher pressure to regions of lower pressure.

Both the direction and the speed of the wind vary with altitude. The wind direction will commonly spiral with altitude, in an effect known as the Ekman spiral, after the discoverer.

REGIONAL AIR POLLUTION METEOROLOGY

Regional scale meteorology deals with atmospheric conditions that cover a geographical area ranging from about 10 to 100 miles in diameter. The time periods may be from hours to several days in terms of their effects on air pollutants and air pollutant dispersion. Specific regional effects and conditions include inversions and sea to land breezes.

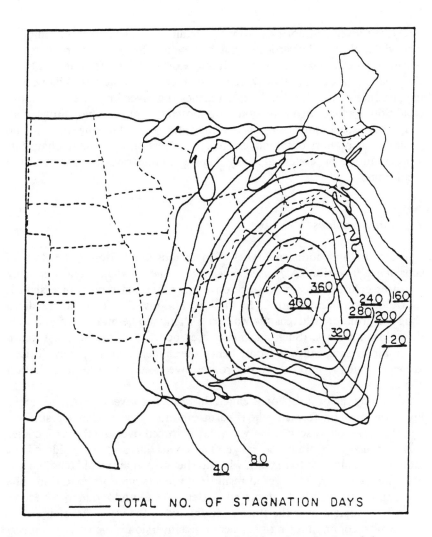

Figure 5.8.   Total number of air stagnation days in episodes of 4 or more days ('36–'65).

Inversions

Probably the most important regional air pollution meteorological effects are those of inversions. This is a situation where, as one moves from the surface to higher altitudes, the temperature increases rather than the normal pattern of temperature decrease. Increasing temperature versus altitude is the *inverse* of normal. Above a certain point, however, the temperature will cool off. The vertical distance in which this temperature inversion occurs is called the *inversion layer*. It may exist on the surface. It may be an elevated condition such that there is a surface layer which exhibits normal temperature conditions before one reaches the inversion base. This latter condition is an elevated inversion, as opposed to a ground-based inversion.

The key aspect to inversions is that they limit vertical mixing (there are no great effects on horizontal movement). An inversion will tend to change its altitude during the course of a day. Figure 5.9 illustrates various inversion conditions.

Types of Inversions

There are three additional types of inversions in addition to the classical subsidence inversion. A *frontal inversion* is one in which rapidly moving warm or cold air masses advance on a more stationary mass of a significantly different temperature. In these conditions, the colder air tends to "hug the ground," whereas the warmer air tends to be buoyed up and over the colder air mass. (Such an impact area between a cold air mass and a warm air mass is called a front, hence the name.) Unfortunately, little is known about the importance of frontal inversions for local air pollution conditions.

A second type of inversion is called the *advective inversion*. An advective inversion may be formed when warm air moves over a colder surface. Another example is seen when warm air is forced over the top of a cooler surface, such as a mountain range. The cooler air on the lee side of the mountain would tend to remain, whereas the warmer air would tend to flow over the cooler air. This would result in an *inversion aloft* condition. The condition is frequently encountered in Denver, Colorado, due to advection and air movement over the Rocky Mountains to the west of the city.

The most common form of surface-based inversion is the *radiation inversion*. This occurs when the surface of the earth has become cooler during the night by the loss of radiant energy. The decreasing temperature of the ground's surface causes the air in contact with it to also lose heat. Thus, there is a cold air mass at ground level. This type of inversion is most pronounced during the late night and early morning hours, and is sometimes called a nocturnal inversion. As the sun comes up, there is heating from below, and an erosion of this inversion layer. Diffusion and vertical motions in the mixing layer then become possible. An inversion layer may then result.

Elevation of an inversion layer determines the *mixing height* for those

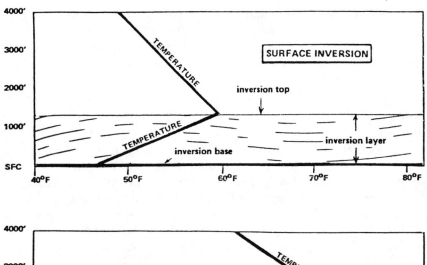

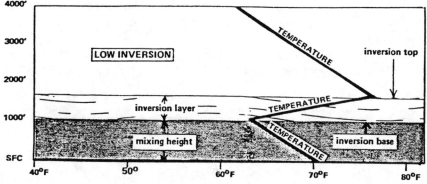

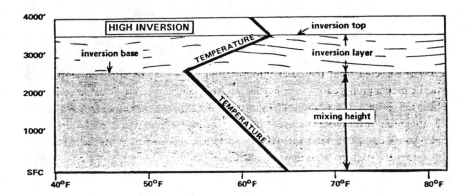

Figure 5.9.   Typical inversions.

contaminants emitted below the inversion layer. It is possible with high daytime surface air temperatures for the inversion to be completely destroyed and near neutral conditions restored throughout the lower troposphere. In these cases, we would find unlimited vertical mixing possible.

The Classic Example

The local effect in California is an extension of the regional effect of the Eastern Pacific high. The inversion-forming processes along the West Coast show a seasonal variation. The classic subsidence inversion is seen in southern California. In summer, the Eastern Pacific high (Figure 5.10) has a pronounced subsidence heating aloft, but as the air mass approaches the California coast, it takes a wide curving path around the northeastern side of the Eastern Pacific high where it moves from colder to warmer parts of the ocean.

The resulting warming of the lower portions of the air produces a convective mixing that distributes moisture evenly through the lowest 100 to 300

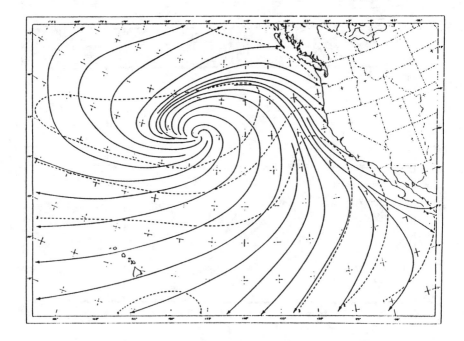

Figure 5.10.   The Eastern Pacific high in July (average position).

meters of the atmosphere and establishes normal stability throughout this "marine" layer. Thus, the base of the inversion does not appear at the surface, but at the top of this shallow marine layer as it has come into equilibrium with the ocean's surface. It is this marine layer, that upon arrival over the Southern California basin, becomes transformed into the "smog" layer.

During the winter seasons, the Eastern Pacific high is much weaker and is positioned much further south. The result is that the inversion along the California coast does not occur so frequently, nor is it as strong. Therefore, occasional cyclonic systems move through the area bringing unstable layers of air, high winds, and precipitation. The disappearance of the smog over Los Angeles (more accurately, "dilution beyond recognition"), is due to the horizontal and vertical ventilation effects of these turbulent higher velocity winds. Rain-out is not a significant factor for air pollutants in these cases.

The high pressure systems which reestablish themselves over the western United States and the Great Basin following the cyclones tend to stagnate, which brings to Los Angeles a low and strong inversion. Fortunately, it also brings the famed Santa Ana winds out of the northeast, which bring an enhanced horizontal ventilation. This prevents the accumulation of high air pollutant concentrations below the inversion. The stronger the pressure gradients, the higher the wind velocities, so that air contaminants are effectively flushed out to sea. The difficulty arises when the anticyclone weakens and the normal flow of air is reversed, bringing the contaminants onshore once again.

As the cells weaken, the wind vectors are so weak that the normal sea breeze develops during the warmest hours of the day, reversing and reestablishing the normal direction of flow causing literally a wall of air contaminants to come onshore. In fact, the situation immediately following the weakening of high pressure cells causes the highest concentrations of photochemical contaminants to move back over land. In this case, however, the air mass trajectory is to the southeast. Therefore, contaminants emitted from Los Angeles tend to arrive in San Diego, having fully reacted out over the ocean for several days before coming onshore. This is a prime example of regional transport of air contaminants.

Sea to Land Breezes

There are other influences which affect how the air moves, and therefore impact air quality. One of these results from the differences between bodies of water and adjacent land masses.

Large bodies of water, such as oceans or large lakes (e.g., Lake Michigan) have significant heat capacity compared to adjacent land surfaces. Also, the surface of the water body will have little temperature change during the course of a day. The sun's radiation will penetrate to depths of 10 to 30 feet below the surface and thus be absorbed by large quantities of water. Currents and eddies will further distribute the radiant energy to their depths. As a consequence, the solar energy will be distributed over a large mass with a high heat capacity.

On the dry land, however, there is no mechanism for advective transport of the sun's radiation. Therefore, all of the radiation is absorbed in the upper several inches of the ground surface. The soil surface temperature rises significantly compared to the temperature of an adjacent large body of water. During the course of the day, the air immediately above the land's surface will receive energy both from convection of air parcels next to the earth's surface and by radiation of heat away from the earth's surface. The air, being heated from below, will tend to rise and form a thermal vent.

The air over the body of water, being cooler and denser, will then move onshore, displacing the warmer and less dense hot air. Thus, any contaminants in the atmosphere over the coast will be vented up vertically. As the air moves up, it will cool and move laterally. Meanwhile, the cooler, more dense air over the water surface will move inland to displace the less dense warm air. It will be replaced over the water by sinking masses of cooled air originally coming from over the coastal plain. This will set up a localized convective cell. During the day, it lofts parcels from the coastal areas up, out, and over the body of water. The same parcels of air then will move onshore in a convective cell. Figure 5.11 illustrates such an onshore cell.

During the nighttime, the opposite effect occurs. Figure 5.12 illustrates

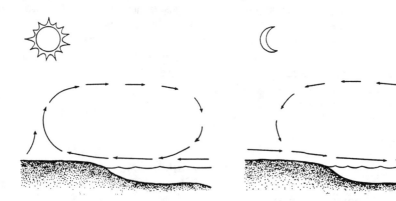

Figure 5.11.   Sea and lake breeze.   Figure 5.12.   Land breeze.

such an offshore land breeze. Particularly on clear cloudless nights, the earth's surface will radiate heat away much more efficiently than the water. Consequently, the surface of the earth will cool. The air masses over water will maintain temperatures closer to that of the water body. If the air mass over the coastal region is cooler, it will become more dense than the relatively hotter area over the water body. There will then be a reverse convective cell set up where the relatively warmer air parcels over the water body will be lofted up and cooled. They will begin sinking over the land mass, while the air originally over the land will tend to move offshore.

These convective cells may extend to heights of one kilometer. In the tropics, these cell heights may extend to three or four kilometers. It is not uncommon for such a convective cell to extend for 10 to 15 miles to either side of the coastal region, although in some circumstances the sea breeze may extend for up to 50 miles.

In general, the offshore land breeze is weaker than the daytime onshore sea breeze. If a radiation inversion is present, the amount of vertical mixing is limited even further, which may lead to severely limited opportunities for air mixing. This increases the potential for greater concentrations of air pollutants during the evenings. Thus, nonphotochemical air pollutants would tend to build up during the evening hours due to such an effect along coastal regions.

Sea-land breezes and convective cells are common in large coastal metropolitan areas such as Chicago and Los Angeles. Such conditions become aggravated when pressure regions form elevated inversion layers, thus capping the convective cell.

Other Dispersive Characteristics of the Atmosphere

Other effects may occur in conjunction with radiation by sunlight which causes heating or cooling of air parcels. Any mechanism which causes air parcels to heat or to cool, thus changing their densities, will cause vertical and horizontal movement of air parcels. Topographic effects will have significant influences on air contaminant levels in areas located next to mountains, river valleys, coastal areas, and lake shore areas.

Valley Effects

Air is a fluid, like water. With valley topography, the air in the valley tends to be channeled in much the same way that water is. Valley walls will mechanically channel the wind into either an "up the valley" or "down the valley" flow pattern. The direction depends on a number of factors including the relative pressure gradients at either end of the valley.

Being confined on two sides, a valley is subject to much more intricate patterns of air movement by virtue of limited horizontal dimensions and the effects of soil or ground cover on either side of the valley. Also, a phenomenon called *entrainment* will cause parcels of air up to 10 meters above a flowing river to move in the same direction as the water, regardless of the general valley wind flow. This has significant implications for catastrophic releases of hazardous air pollutant chemicals on or into such a river.

Apart from these influences, colder and denser air tends to drain down the valley to a lower elevation. Therefore, it is not uncommon to find that the local air movement in a valley will be different from the synoptic scale or regional meteorological conditions. Figure 5.13 illustrates nighttime wind patterns.

Those parcels of air closest to the river (influenced by the relative proximity and height of the mountains forming the valley) will be cooler and therefore more dense than the air at higher elevations in the same valley. Thus, on a local scale, the influence of temperature will tend to maintain air flow directly above the river in the same direction as the river.

Depending on the orientation of the river valley, i.e., east/west versus north/south, differential ground heating effects also substantially influence air flow patterns. During nighttime hours, when heat is lost at all portions of the valley, lower air parcels will be cooled by radiation loss and thus become more dense. These air parcels will move "down slope" to the valley floor, causing a bulk flow of air down the valley.

When the sun rises on a north/south-oriented valley, one side of the canyon or river valley will be heated while the other will still be relatively cold. This case would form an overturning or spiraling flow pattern, where

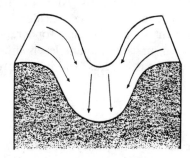

Figure 5.13.   Flow of slope and valley winds at night.

the air on one side of the valley will be heated and begin rising, while the cold air masses will still continue to downslope. This situation would remain until the entire surface of the valley is in direct sunlight, at which point the air would begin to move vertically.

In east/west-oriented valleys, such overturning flow patterns would not tend to occur as dramatically. Since the sun is in the southern portion of the sky (in the Northern Hemisphere) this effect would occur during all daylight hours. The effect would be moderated by the sun angle as a result of the season.

A common occurrence is the formation of a nocturnal temperature inversion in a river valley. In the evenings, cooler, dense air from the slopes tends to accumulate, effectively filling the valley with colder, more dense air which intensifies the surface inversion that would normally be produced by loss of heat through radiation. This inversion deepens over the course of the evening and may reach its maximum depth just before sunrise. The height of the inversion layer depends on the depth of the valley and intensity of the radiation cooling. Any contaminants emitted into this valley, therefore, would tend to remain there until such time as either the chimney effect occurs, or the inversion is broken by surface heating.

Depending upon the relative temperatures and intensities of the inversion, the inversion itself may or may not be broken. In that event, valley regions may experience very limited vertical movement, thus having a situation similar to a stagnation in which contaminants build up. The only relief in such situations may be the movement of the anticyclone away from the area, (causing loss of the inversion) or limited relief due to the downslope nighttime winds carrying some contaminants farther down the valley. Figure 5.14 illustrates such a condition.

Chimney Effect

The "chimney effect" is significant in moving air next to a slope vertically. In this situation, under sunny conditions, parcels of air heat faster and become less dense along a mountain ridge overlooking a valley or plain. The bulk of the air in the lower elevations would tend to remain at the same temperatures. The air parcels next to the slopes would move up and over the ridge such that even if an inversion were present, this inversion would be broken along the 10 to 50 meter thick boundary layer adjacent to the slope. Air masses from below would then be lofted up and out from the valley or plain.

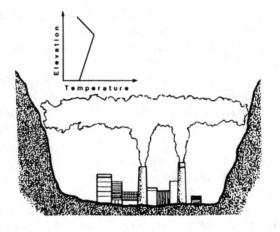

Figure 5.14.   Nocturnal inversions in a valley.

Vegetation Effects

Should slopes in any terrain be covered with vegetation, the slope winds may be modified, since vegetation such as trees and forests do not allow the ground surface to reach high temperatures during the day. In this situation, the vegetation absorbs radiation and gives off moisture through evapotranspiration. Vegetation, therefore, may significantly alter slope winds and essentially form a boundary layer at ground level.

Recent studies indicate that a typical forest effectively forms a canopy, trapping a layer of air below the treetops. The contaminant levels experienced in the breathing zone of people in those areas may significantly change from that assumed to occur in a terrain without vegetation.

Mountain Effects

The influences of a mountain range may have the effect of causing air contaminant buildup. When air flows across a mountain range, a situation similar to the chimney effect occurs during the daytime. In the evening the reverse occurs with downslope winds. The cyclic nature of such mountain winds can encourage pollution episodes. Denver is in such a position, with a mountain range trending north/south to the west of the city with essentially plains to the east.

Apart from the effects of weather fronts moving through the area, daily wind regimes in Denver are established in which downslope drainage at night with early morning winds carry air contaminants from the city toward the northeast. The flow reverses during the afternoon and easterly winds carry the polluted air masses back across the city. The consequence of heavy air emissions in the central city and the northeast often takes place in the late afternoon or early evening. Also, radiant heat effects during the night set up an inversion over the city which effectively acts as a cap, thus further intensifying air contaminant levels.

Urban Heat Island Effects

Another effect which is commonly experienced in metropolitan areas is the Urban Heat Island effect. In this effect, the built-up portions of metropolitan areas, commonly with high-rise buildings, are found to be several degrees warmer, on the average, than the surrounding countryside. This is due to the large surface area and heat capacity of the concrete and asphalt that are used in typical metropolitan areas.

The absorption of radiant energy by these materials causes a heat buildup. As a consequence, air blowing into these metropolitan areas tends to rise by picking up additional heat. This sets up a recirculation cell over the city, similar to the convective cells noted earlier under the land-sea breeze situations. As a result, few surface-based inversions occur over urban heat islands. However, elevated nocturnal aversions of longer duration may occur frequently.

Such a recirculation cell, as seen in Figure 5.15, tends to maintain polluted air masses over an urban area. This, associated with high concentra-

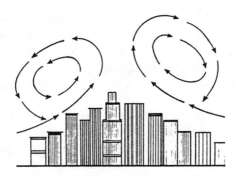

Figure 5.15.   The urban island heat effect.

tions of automobiles and other local sources, tends to give high visibility to those urban areas from observers located miles away. These "domes" of recirculated contaminants can build up and become quite visible when other portions of the atmosphere are clear.

Horizontal and Vertical Air Patterns

Both horizontal and vertical air flow patterns are critical when it comes to understanding the dispersion of air contaminants. Each may significantly influence options for managing air quality. There is little that can be done to modify meteorology.

Atmospheric Stability

A key concept in understanding mixing of the atmosphere, particularly in the vertical dimension, is that of the *atmospheric stability*. This is a measure of the buoyancy of a given parcel of air and is not confined to the boundary layer of the atmosphere (the lower several hundred feet), but will be effective throughout the lower troposphere. The *lapse rate* is the key empirical method of determining the stability of the atmosphere with respect to vertical air parcel movements.

It is observed that a parcel of dry air will cool spontaneously at a rate of approximately 1 degree centigrade for each one hundred meters of vertical rise in altitude. This is the basis for determining a reference point for atmospheric stabilities. This temperature decrease versus height is seen in Figure 5.16 as the dashed line. It is termed the *dry adiabatic lapse rate*, and is considered the neutral point for atmospheric stability.

The actual temperature profiles versus altitude for a given location will vary, as will their respective stabilities. If an air parcel shows a lapse rate in excess of -1°C/100 meters ("-" denoting a temperature *decrease* with increasing altitude), we consider it to be an unstable atmosphere. It is possible to have a "super adiabatic lapse rate," which is an unstable atmosphere where the lapse rate is in excess of the -1°C/100 meter reference point (curve A in Figure 5.16).

If the atmospheric condition is equal to the adiabatic lapse rate, we consider that to be a neutral atmosphere. As the lapse rate changes to values between 0 and -1°C/100m of altitude (i.e., becomes less negative), the atmosphere becomes progressively more stable (curve B). If one finds that the layer of air being considered is at the same temperature at all heights, it is an isothermal condition (curve D).

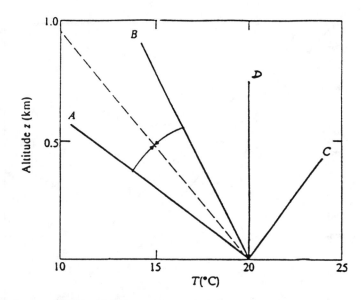

Figure 5.16.   Lapse rates: air temperature vs. height.

On the other side of the isothermal lapse rate, we would find a positive increase in temperature with altitude. Thus we might have a + 1.0°C/100m rate of increase (curve C in Figure 5.16). Being in the positive direction, we consider this to be an *inversion* condition, compared to the dry adiabatic lapse rate. This condition has significant impact on air pollutant dispersion.

An actual temperature profile showing temperature versus altitude for various periods of time under the influence of a radiation inversion is shown in Figure 5.17. The reference line is the *dry adiabatic lapse rate* shown in the figure. One may move from the early morning hours, where the inversion is clearly present, to the morning and early afternoon hours where the inversion is gone, back to a reformed ground-based inversion. This condition severely limits the vertical mixing capacity of the atmosphere.

Vertical Mixing

The movement of air in the vertical dimension is enhanced by major temperature differences. The greater the temperature difference between the surface and higher elevations (i.e., the temperature gradient), the more vigorous the convective and turbulent mixing of the atmosphere will be.

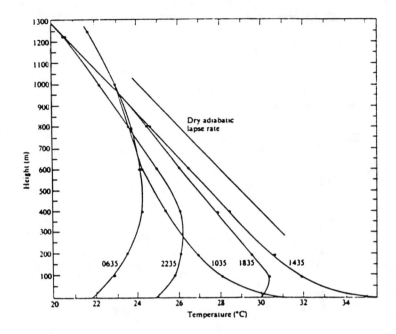

Figure 5.17.   Temperature profiles showing the erosion and formation of a radiation inversion.

Likewise, the greater the area of the vertical column over which turbulent mixing occurs, the more effective the dispersion process will be.

The Maximum Mixing Height (MMH) is the height of the convective layer associated with the maximum surface temperature. The MMH exhibits both local and seasonal variations. In addition, it is affected significantly by topography and large-scale air movements. During the day, the minimum value will occur typically just before sunrise. As the sun heats up the ground's surface, the air next to it will expand, become less dense and begin rising by convection. Cooler air from aloft descends to take its place and a recirculation pattern appears. As solar insolation continues, the vertical motions will become more intense and the *maximum mixing height* will be reached usually in the early afternoon, with values in the thousands of feet.

Minimum values will be observed late at night or in the early morning hours and may be the result of a surface based inversion noted earlier. In these cases, the mixing height may be near zero. Whatever the MMH, it will represent the vertical depth (and therefore, volume) in which air contaminants may be mixed. Topographic features such as water surfaces will result in lower mixing heights due to their large heat-absorption capacity. On the other hand, bare ground surfaces such as deserts may have maximum mixing heights of over 15,000 feet.

Horizontal Air Movements

The direction of the wind and its speed, particularly near ground level, in addition to local processes such as turbulence, have a significant impact on movement and dilution of air pollutants. In the presence of an elevated inversion, the surface wind speed and direction may be the predominant dispersal process occurring.

We call air which is not moving smoothly *turbulent.* Turbulence is described as eddies in a local air parcel which produces mixing by two mechanisms. The first of these is purely *mechanical turbulence,*which is caused by irregular air movement over terrain such as trees or around obstructions such as buildings. This causes mixing of smaller parcels of air with relatively uncontaminated air. Mechanical turbulence serves to enhance dispersion by increasing the eddies around building edges. *Thermal turbulence* occurs as the result of radiation heating from the sun on various objects. Differential surface heating occurs which leads to smaller parcels of air moving upward, mixing with contaminated parcels.

Terrain impacts, or surface roughness, may significantly affect the speed of the wind across a given geographic location. As seen in Figure 5.18, wind speed may be significantly retarded over built-up areas to heights of several hundred meters due to the inertial drag and turbulence induced by building obstructions. Rural areas, of course, do not experience the same degree of inertial drag due to turbulence as in built-up areas. This is also termed a boundary layer effect.

The influence of *wind speed* is to increase the dilution of air contaminants. As an example, if a parcel of air is moving across an emission source at a given wind speed, the emission (with a rate of "Q") would be diluted into a volume represented by the area of the plume times the wind speed per unit time. As illustrated by Figure 5.19, we would find the concentration at some point downwind to be:

$$\text{concentration} = Q/ (\mu \times A) \qquad (5.1)$$

where: Q = emission rate, grams/sec
μ = wind speed, meters/sec
A = plume cross section, square meters

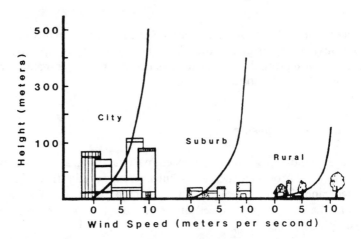

Figure 5.18.   The effect of surface roughness on wind speed.

Thus, if the wind speed is doubled to 2 μ, we would find the dilution (all other factors being equal) to *halve* the down wind concentration. Thus, we see the direct influence of wind speed on down wind concentrations. This model example is greatly complicated in real life by the actual nature of plume dispersion.

LOCAL AIR POLLUTANT DISPERSION

Any action which encourages more effective dilution of air contaminants from a source will lower the contaminant concentration at the ground level. Having a source located on the top of a hill would be expected to enhance its

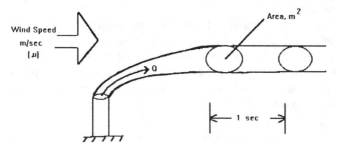

Dilution of Air Emission Rate "Q"

Figure 5.19.   Dilution of air emission rate "Q".

dispersion. Likewise, tall chimneys (the first air pollution control device) serve to increase the volumes of air which will dilute an emission.

Dispersion may be evaluated from two aspects. The first is by type of source: point, line, and area. A point source corresponds to a single exhaust point, such as a chimney or stack in a specific geographical location. A line source represents emissions from a road or a vent from a long building. An area source consists of a number of individual sources so located that they tend to merge into one large mixed source. Landfills and cities may come under the heading of area sources.

The other aspect is by evaluating those conditions which enhance plume rise. *Plume rise* is the additional vertical movement of a plume above its original release height. The shapes and characteristics of various plumes under given atmospheric stability classifications may also be evaluated. These lead to a better understanding of which conditions cause the greater dilution.

Point Sources and Plume Dispersion

Point source emissions of air contaminants have been the most extensively studied and are the best understood of all three source types. Point sources are characterized by the entire emission occurring through a single opening, whether a chimney, vent pipe, or exhaust duct. This point may be at or near ground level (typical of evaporative emissions), or at some elevated point (such as from a power plant stack). Also, they tend to be "powered exhaust" emission sources—the discharge is accomplished by a fan or blower driving the exhaust gases from the generation process through duct work or piping to the chimney. Only a few cases use *natural draft* to release exhaust gases from the stack.

Plume dispersion is enhanced by the highest possible effective stack height. The amount of plume rise that may occur over and above the actual release height (top of the stack) will determine its overall *effective stack height*. In Figure 5.20, the height above the stack at which the plume has attained a completely horizontal component is the plume rise. The effective stack height is the release point height plus the plume rise. Thus, the greater the effective stack height, the greater the dispersion will be because there will be a greater volume of air between the plume and the ground surface.

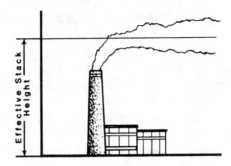

Figure 5.20.   Plume rise schematic.

Plume Rise

Plume rise is generated by two effects: momentum or mechanical effects and temperature. Mechanical effects are noted by the transference of momentum from the stack gases at the release point to the atmosphere. In a typical crosswind situation, the higher the wind speed, the more quickly the plume loses momentum and therefore bends over. The plume rise is therefore decreased with higher wind speeds. On the other hand, with high wind speeds, the plume is diluted to a greater extent.

Turbulent effects are also important for small sources that have relatively small exit velocities with a negligible buoyancy factor (cold exhaust). Velocity is the major component of momentum. To be effective, stack gas velocities must exceed the average crosswind speed by 20% to 30% to prevent entrapment of exhaust gases on the turbulent downwash of the exhaust stack and associated structures. "Cold" releases require a higher chimney or stack to avoid this problem.

Temperature has a greater impact on plume rise. In this situation, the exhaust gas buoyancy, due to the temperature difference between the hot stack gases and the atmosphere, provides for much more effective plume rise. Based on total energy inputs, buoyancy due to temperature is greater by a factor of about 10:1 over the dilution accomplished by momentum. Therefore, the hotter the exhaust gas, the more effective, and greater, the plume rise will be.

In general, therefore, without the influence of an inversion, effective plume rise is directly proportional to the heat release rate and inversely proportional to the wind speed. One empirical formula, among many, fitting these parameters is the simplified CONCAWE formula (Conservation of Clean Air and Water, Western Europe), which states:

$$h_r = k \, Q_h^{1/2} \, \mu^{-3/4} \tag{5.2}$$

where: h_r = effective plume rise, meters
k = empirical constant of 0.086 m²/watt-sec
Q_h = heat release rate, watts
μ = wind speed, meters/sec

A moderately good fit to the observed plume rise is obtained using this expression from some European sources for power plant exhaust plumes.

Plume Shape

The shape of an elevated plume is a strong function of the stability of the atmosphere in which it is released and in which it attains its final effective stack height. It is informative to review these characteristics as a function of the lapse rate.

In Figure 5.21, we see six possible combinations of stability classifications and/or inversion conditions, together with their influence on plume structure. For the release of exhaust gases into an unstable atmospheric condition, or super adiabatic lapse rate (A), the plume will assume a looping condition, due to the extreme instability of the atmosphere. This is most conducive to the dispersion of air contaminants. Yet it should be noted that the plume is more likely to touch the ground's surface closer to the emission point and, therefore, instantaneous concentrations may be significantly above ambient air quality standards.

In the case of an emission into a neutral or slightly stable atmosphere (B), we would see a "coning" condition occur in which the plume disperses fairly equally into both horizontal and vertical dimensions. This is conducive to good dilution for any given downwind distance. The instantaneous shape is influenced significantly by the amount of meander or variation in wind direction, which will also cause the plume to be dispersed and diluted.

Should the release be into an inversion (C), such as a deep surface inversion, there will be no tendency for the plume to move up or down. Therefore, a fanning plume would be seen as a direct function of wind meander.

Where the release height is above a shallow surface inversion into a layer where we have neutral or slightly stable conditions above the inversion, the plume would have a tendency to continue rising due to buoyancy effects

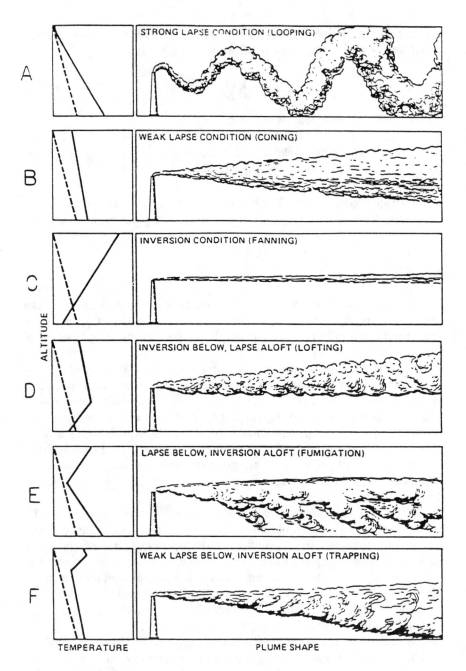

Figure 5.21.   Plume structure as a function of temperature/stability.

and natural turbulence. It would remain isolated from the ground surface due to the presence of the inversion below the release height (see Figure 5.21D).

With an inversion above the plume rise and an unstable atmospheric condition below we would see a plume shape similar to Figure 5.21E. Depending upon the relative height of the inversion base above the release point, we may get a *fumigation* effect, whereby the plume effectively bounces off the base of the inversion and is reflected downward through the unstable layer. The plume would fumigate the surface at fairly close distances to the release point.

There are a few situations, however, where the stack height is sufficiently close to the inversion base and the heat release rate (Q_h) is sufficiently high that the plume can literally punch through the inversion. For a given source, this is termed the *critical inversion height*. The critical inversion height is a function of the strength and depth of the inversion, stack height, exhaust gas temperature, heat release rate, and other variables. It is not a constant, and typically is useful only with very low inversions and high heat release sources such as power plant stacks.

Another condition would be one with an elevated inversion layer above the effective stack height with a neutral to slightly stable atmosphere below the inversion base. Figure 5.21F illustrates such a condition in which *trapping* would occur. In this condition, the plume would reflect off the base of the inversion and be fully dispersed in a fashion somewhat similar to the coning plume, only it would be totally contained below the inversion base. The possibilities for a critical inversion height existing here are better in this case than for Figure 5.21E.

Line Sources

A line source such as a road is somewhat more complicated than a point source due to the fact that a road is continuous and there are extended portions of the line which must be considered. Likewise, the wind speed and direction with respect to the line influence the downwind air concentrations. Also, there is no emission into atmosphere from an elevated chimney or stack. These are surface emissions with very little buoyancy.

The plume shape for a line source is therefore nearly impossible to describe, based upon simple equations. Therefore, numerical techniques must be used with high-speed computers.

Vertical *diffusion* and crosswind dilution are the major plume features in a line source. Vertical diffusion processes predominate in these equations. Therefore, we have the highest concentrations right at the boundary of the road, and they "decay" as one moves away from the line source. A theoretical representation of a line source plume's relative concentration is seen in Figure 5.22.

Area Sources

Area source emissions may occur from a relatively large source such as a city or from one relatively small, such as a landfill. Effects of horizontal diffusion, geometry of the sources, and varying overlapping plumes and/or release locations, together with variations in operations make area sources exceedingly difficult to evaluate.

In general, for regulatory modeling purposes, such an area is assumed to be a large point source with a low plume rise. Inversions aloft further decrease the potential for area source dispersion in the vertical dimension. Diffusion and wind speed are the dominant parameters for area sources. Figure 5.23 shows such a projection of an area source back to a single point source typically used in dispersion modeling.

DISPERSION MODELING

In air dispersion modeling, contaminant concentrations are calculated at some point in time or space based on our knowledge of fluid flow, source emissions, meteorology, and past data. Modeling may be performed by the use of physical devices or calculated using mathematical functions.

In *physical modeling, scale models* of the terrain, buildings, dimensions, and relative distances in question are constructed and immersed in a moving fluid in fluid flow tanks or wind tunnels. A visible material, either hydrogen bubbles or smoke trails, is released from different heights or locations in the model to evaluate the movement and dispersion of fluids around the physical obstructions. Recording devices such as video cameras are used to document the flow fields and diminution of visible intensity (taken as proportional to concentration). These are usually accompanied by micro scale temperature, velocity, and pressure measurements.

From these physical models, empirical relationships are determined between shapes, locations, relative dimensions and the fluid parameters: flow, velocity, etc. These relationships are then applied to a specific source for purposes of advanced planning, for dispersion design around structures, locations of stacks, and maximum stack heights. The scale covered may be an area from several city blocks to a mile or more in diameter.

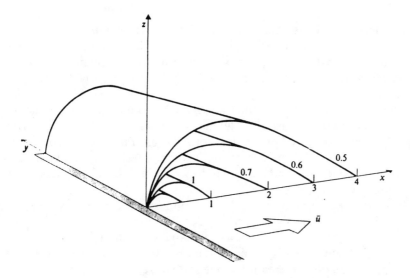

Figure 5.22.   Line source relative cross wind concentration.

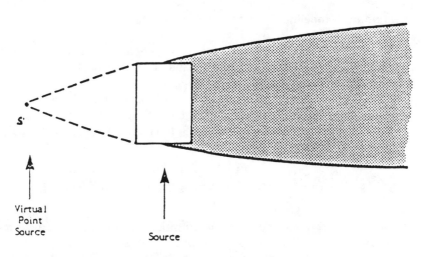

Figure 5.23.   Virtual source approach for line and area sources.

Mathematical models can be either *deterministic* or *statistical* in nature. The former deal with modeling air contaminant dispersion as a function of known scientific laws or relationships in order to determine a downwind concentration of a contaminant based upon its source strength, temperature, etc.

Statistical models, on the other hand, are based upon empirical relationships between known past air pollutant concentrations (usually ozone) and specific meteorology. Mathematical formulas are then utilized to evaluate the influence of different meteorological conditions on contaminant concentrations which have already been dispersed into the atmosphere or will be formed photochemically.

The main reason mathematical models are used is that it makes possible the evaluation of various scenarios under different atmospheric conditions and stack heights with less cost, as compared to fluid models or construction of field equipment with air monitoring. It is anticipated that the use and application of mathematical models will increase in importance as long-term air quality management strategies take effect.

Deterministic models are those which have been derived from observations of "real life events" and are modified by our knowledge of basic science, engineering, and meteorology. In general, these models are mathematic algorithms run on computers and are used to calculate a downwind concentration along a plume centerline based on a Gaussian distribution.

Point Source Modeling

The generalized Gaussian plume model for a point source utilizes dispersion along the direction of an average wind vector to calculate the ground level pollutant concentration.

In the Gaussian model, the spread of an elevated plume in vertical and horizontal directions is assumed to occur by diffusion along the direction of the mean wind (Figure 5.24). The ground level concentration is calculated by means of Equation 5.3.

$$c(x) = \frac{Q}{2\pi \, \mu \, \sigma_y \, \sigma_z} \exp\{-0.5(H/\sigma_z)^2\} \exp\{-0.5(y/\sigma_y)^2\} \qquad (5.3)$$

where: c = ground level concentration (g/m^3) at some downwind
             distance "x" in meters
       Q = average emission rate (g/sec)
       μ = mean wind speed (m/sec)
       H = effective stack height (m)

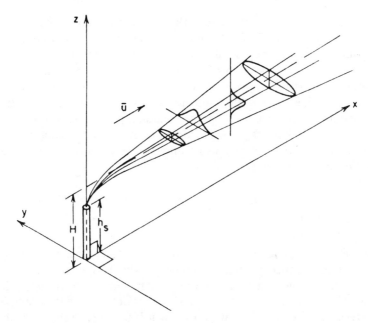

Figure 5.24.   Gaussian distribution of pollutant concentrations across the plume.

σ_y = standard deviation of wind direction in the horizontal (m)

σ_z = standard deviation of wind direction in the vertical (m)

y = off-centerline distance (m)

exp = natural log base (equal to 2.71828) raised to the power of the bracketed values.

H includes the physical stack height plus the plume rise (Figure 5.20). The plume rise is calculated by Equation 5.2 or other techniques. The parameters σ_y and σ_z are *dispersion coefficients* (empirical factors) used to determine horizontal and vertical dispersion. They predict plume spread horizontally and vertically for given distances downwind under different conditions of atmospheric stability. These vary depending on whether the source is located in an urban or rural setting.

A schematic of the Gaussian plume model in three dimensions for a point source emission is illustrated in Figure 5.24. This model is limited to point sources with an emission release point *above* the downwind receptor points.

Due to wind meander, the plume itself will wander over the ground surface. Therefore, the calculated ambient concentration of pollutants downwind will fluctuate with time as the turbulence changes the directional position of the plume. Thus, the hourly average at any given point will be lower than an instantaneous value due to the plume meander.

The surface concentration profile is seen in Figure 5.25, which illustrates the Gaussian distribution of concentration along and to either side of the plume's centerline. The vertical dimension is the relative concentration at respective downwind points. It is the Gaussian plume models which have been used for the longest period of time and have EPA regulatory status. Model accuracy varies from plus or minus 30% to plus or minus 200%, depending on the circumstances.

Model Averaging Time

The current approach in dispersion analysis is to characterize downwind air contaminant concentrations by long-term (hourly) averages. The long-term average plume behavior is more regular and therefore more simply described in mathematical terms. Our science is not sufficient to completely characterize plume behavior. We do know it is turbulence that is primarily responsible for dispersion rather than molecular diffusion.

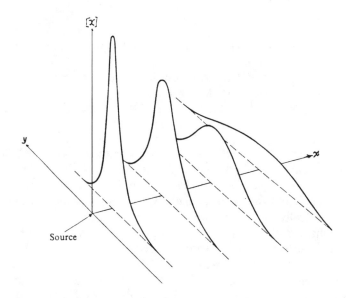

Figure 5.25.   Gaussian distribution of surface pollutant concentrations downwind of a point source.

All of these models assume a steady state condition. In other words, we must assume that for certain averaging periods (typically one hour), emission rates, wind vectors, and atmospheric stability are a constant. We may then derive a concentration field at the surface *receptor* downwind of the *source* emission point for the hourly average. For longer averaging periods such as 24 hour averages, or annual averages, we use an average of individual one-hour averages to derive the longer term average concentrations at various downwind points.

Plume Model Modifications

Mathematically, we may treat a plume from a point source as entraining a three-dimensional volume until it reaches the ground. Once this occurs, the contaminant spread does not continue since there is no flux of plume gases into the soil. Mathematically, we treat this enhancement of the downwind concentrations inside that theoretical volume as being increased by the amount of volume not penetrating through the surface of the soil. This assumes a reflection of the air contaminants back up from the surface of the earth. This flux gives us a virtual image source identical to the original source by this theoretical reflection. We take downwind surface concentrations and mathematically enhance it for this image "source." Figure 5.26 illustrates this approach.

Other modifications include wind speed. Wind speeds aloft may increase by 30% to 50% over ground speeds. Therefore, we accommodate the winds aloft by an enhancement factor which depends upon the stability of the atmosphere. These enhanced wind speeds over ground level speeds are utilized in the Gaussian plume model equations.

The location of that maximum impact point depends on the wind speed and the atmospheric stability. Consequently, under steady state conditions we would find that point to vary from one to ten kilometers downwind from a typical power plant stack of 300 feet.

This maximum point is used in many of the conservative risk assessments currently in use. Risk assessments, as we have seen earlier, depend on a downwind concentration for the inhalation factors of a hazardous air pollutant. In a 2-dimensional presentation of the surface concentrations we will find a series of concentric isopleths of concentration centered around the point of maximum impact.

As the stability of the atmosphere increases from a super-adiabatic to neutral to isothermal to high stability classifications, the downwind distance to the point of maximum impact will increase. The concentration will decrease, however.

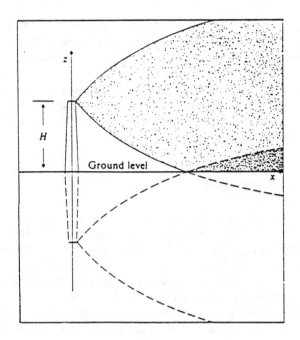

Figure 5.26.   Virtual image source by ground reflection.

A major area of concern is where a plume may encounter an elevated receptor such as a hill or mountain. Models take into account these obstructions and allow for mathematical manipulations of the plume centerline to more closely approximate the surface concentrations at elevated receptors.

Line Source Models

The line source models are handled as a continuous sequence of individual point sources over a near infinite length of point sources. In general, we solve partial differential equations for line sources using the source strength *per unit length*.

The concentrations are affected by wind speed (inversely) and directly as a function of exponential decay. Being ground level releases, the release point is at the surface.

The location of maximum impact will be immediately adjacent to the line source. This is assuming a perpendicular crosswind. As the wind changes to a flatter angle, the concentrations will be increased. The concentration downwind will increase until the wind is virtually parallel to the line source, at which point the maximum concentration will be significantly higher than in a perpendicular crosswind.

Area Modeling

Area sources, apart from statistical models or regional models for planning purposes, use the virtual source approach. In this approach (seen in Figure 5.23), we take the average wind direction and project it back geometrically to an equivalent single point source through which we mathematically "force" the emissions from that area source. We then calculate downwind concentrations from the edge of the original source. For urban areas, these virtual source models are rapidly being replaced by planning models.

Catastrophic Releases

Standard dispersion modeling approaches are not readily applicable in the case of a catastrophic release such as an explosion, fire, or tank car spill. Typically, these accidental releases are analyzed by using vapor cloud dispersion models. These represent significantly different conditions.

River spills and tank ruptures require much different approaches for the source emission rate (which may be varying with time over a very short period), and the plume rise—if any. Indeed, the plume may have a negative buoyancy due to temperature and density effects, and will most probably be a ground level release.

Likewise, momentum effects may be significantly more important in catastrophic ground level releases than in elevated continuous releases. They are further complicated by the fact that many of these catastrophic releases begin as a liquid spill such as from a pipeline rupture. Therefore, two phases may be present initially—a liquid jet and a vapor cloud. Prediction models for these scenarios are not conclusively verified. Catastrophic releases may drastically change the micro scale environment, and therefore usual assumptions about bulk gas fluid properties (approximating air) may not be valid.

Approaches taken in these models are numerical simulations typically involving a dense gas model. A number of privately developed models are available which attempt to simulate such releases. The significant features of a catastrophic release are shown in Figure 5.27.

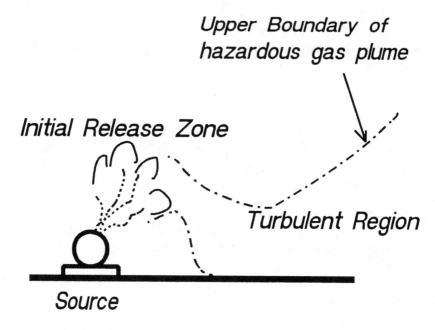

Figure 5.27.   Hazardous gas modeling regions.

PLANNING MODELS

The latest approach in modeling area sources, particularly urban areas, is to evaluate air quality impacts based on non-Gaussian methods. In these planning models, an urban area is computer-modeled utilizing finite element or Eulerian methods. The geographic area is a grid-based three-dimensional cell which extends to the base of the inversion (if any), or to the maximum mixing height. These models are typically used for long-range strategic planning and are mandated by the 1990 amendments to the Clean Air Act. To date, the major focus has been with attempts to simulate ozone concentrations over a large area.

In these grid-based models, we have small boxes or cells which are mathematically distinct *through which* contaminants enter and leave as a function of wind speed, diffusion processes, etc. Prognostic models provide solutions through numerical calculations over one, two, or three dimensions. The photochemical air quality models fall into this category. The EPA's Urban Airshed Model is a grid-based photochemical air quality simulation. In these models, photochemical reactions are allowed to occur within the cells over a period of time. Figure 5.28 illustrates such a cell or box element. The objective is to model the maximum one-hour ozone concentration at the surface of each grid cell. For formulating state implementation plans (q.v.) to achieve the national ambient air quality standard for ozone, the EPA requires use of the Urban Airshed Model.

The simplest of the numerical models are the box models. In this approach, the entire region to be modeled is a group of three-dimensional boxes with the top set at the base of the inversion (if any), or at the maximum mixing height. Emissions are assumed to be distributed instantaneously over the entire volume of the box, and homogeneous throughout. Ventilation characteristics of the model are represented by specifying an average wind speed and rate of rise of the upper boundary. Pollutant concentrations are described by a balance among the rates at which they are transported in and out of the box, their rates of emission per unit time, rates at which the box volume expands or contracts, and the rates at which the pollutants are formed, destroyed, and\or flow out of the box.

Trajectory models represent other approaches based on atmospheric dif-

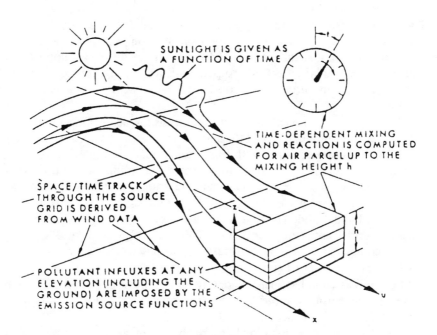

Figure 5.28.   Lagrangian box model simulation schematic.

fusion equations, but they use a moving coordinate system to describe pollutant transport. A hypothetical column of air is defined as bounded on the bottom by the ground and on the top by the inversion base (if it exists), versus time. Given a specified starting location, the column "moves" (by the computer) under the influence of the prevailing wind vectors, passing over emission sources which emit pollutants into the column. Chemical reactions may be simulated in that column. One trajectory model is called the Empirical Kinetic Modeling Approach (EKMA). This is a city specific, or region specific model. It was developed to relate levels of peak ozone to levels of reactive nonmethane hydrocarbons and oxides of nitrogen.

Parameters which may be varied and which are modeled include: wind vectors, photolysis rates, mixing heights, dry deposition, water vapor concentration fields, vertical eddy diffusion coefficients, and temperatures. These models will become more significant as air quality management strategies are evaluated for ozone containment areas of the nation.

STATISTICAL AIR QUALITY MODELS

The need for an effective method of determining one to two days in advance the potential for air pollution *episode* conditions has given rise to the advent of statistical models. These models are not based upon fundamental scientific laws but rather upon statistical correlations of measured contaminant concentrations (ozone, etc.) with measured meteorological parameters.

These statistical models are used for forecasting acute concentrations of ozone, carbon monoxide, PM10, and sulfur dioxide. Within a given geographical location equipped with complete air quality monitoring and meteorological data, it is possible to forecast maximum concentrations to accuracies within ±10% of measured values up to 24 hours in advance.

In order to meet this level of accuracy, certain assumptions must be made and found to be valid. The first is that there is a constant emission rate over the geographic area being considered. Typically, this would be valid over an entire complete air basin where drastic emission changes are not occurring within short periods of time.

Secondly, a complete meteorological data set must be present. There are over 100 meteorological parameters which may enter into the evaluation of such a statistical forecasting model.

Finally, there must be a "calibration point" at air monitoring stations measuring air pollutant concentrations alongside the complete meteorologi-

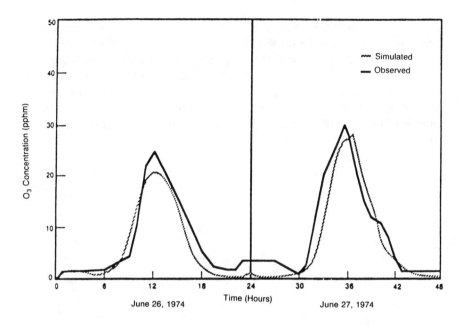

Figure 5.29. **Simulation of the ozone concentration in the Los Angeles area.**

cal data set. This allows for the modification of the statistical model equations to "best fit" the measured condition at the calibration period to some period in the past.

In general, these statistical forecast models take the measured contaminant concentrations for today, along with the complete body of meteorological conditions existing today, and modify them with factors which will yield the maximum concentrations anticipated for tomorrow. These modifying factors include day of the week, season, and year, as well as the monitored values for today.

Modeling efforts used to simulate the anticipated conditions for the following day are found to be reasonably reliable for forecasting maximum levels and acute health impact levels. Figure 5.29 illustrates a simulation of ozone concentrations at one station in Los Angeles, and compares them to the monitored ozone concentrations.

Some of the meteorological conditions which provide input to these models include the upper atmospheric weather, lapse rate, local winds, pressure gradient over the area, persistence of existing synoptic conditions, humidity, inversion structure, the coefficient of scatter, dew point depression, and an autocorrelation factor.

The values and correction factors for each of these are *very specific to the location* in which the statistical models have been developed. Also, all models developed for a specific region will ultimately over-forecast the maximum concentration, due to the general lowering of air contaminant emissions as a result of the implementation of air quality management strategies.

6   STATIONARY SOURCE CONTROL APPROACHES

There is no fire without some smoke.

John Heywood, *Proverbs*, 1546

Probably nowhere else has the impact of air quality management strategies been seen more clearly than in stationary source emissions control. This is not unexpected since historically, stationary sources have been the major anthropogenic emission source category. However, there are a limited number of "back-end" technologies that may be used to control emissions from stationary industrial and commercial processes. As a result, *management options* which reduce air emissions have come into play in addition to installation of air pollution control hardware.

The major new areas of stationary source emissions reduction have been in *source reduction, planning, and design modifications*. Each of these are evaluated in the following sections followed by discussions and evaluations of current "back-end" control technologies.

SOURCE REDUCTION

The most cost-effective approaches to controlling air contaminants are those that entail source reduction. There are four major source reduction approaches. Each requires an in-depth understanding of the processes and activities that emit air contaminants. These include:

- management and operational changes
- process optimization
- combustion modifications
- fuel modifications

Each of these approaches has a different impact on criteria air pollutants, volatile organic compound (VOC) emissions, and nonorganic hazardous air pollutants. VOCs are of concern both as precursors to ozone formation and for their potentially hazardous or toxic properties. Table 6.1 summarizes the relative impact of source reduction approaches on VOCs, hazardous air pollutants, particulate matter, NO_x, CO, and SO_2.

MANAGEMENT AND OPERATIONAL CHANGES

Since *source reduction* approaches have the greatest potential for an immediate reduction in air contaminant emissions, management audits and inventories of both materials and processes are the first steps in defining appropriate options.

Auditing is a process where activities are identified and evaluated for their potential to emit various air contaminants. Environmental audits generally outline *where* and *in what amounts* air contaminants are being emitted. From these audits a plant-wide inventory or profile is generated which identifies activities and contaminant-generating processes with an outline of approaches to mitigate those emissions. Specific followup includes implementing options for control by hardware, minimization by operational changes or product reformulation, and cost-benefit analyses of the various approaches.

Along with audits and inventories, the first action is generally to "clean up the shop." This involves ensuring that spills are minimized, that process equipment is operated properly, and that personnel are thoroughly *trained* in good housekeeping practices. It should be noted that the definitions of

Table 6.1. Impact of Source Reduction on Air Pollutants

Technique	VOCs	HAPs	PM	NO_x	CO	SO_2
Management operations	+ +	+ + +	+ +			
Process optimization	+ +	+ +	+			
Combustion modifications	+ /-	+ /-		+ + +	+ /-	
Fuel beneficiation	+	+ /-	+ +	+ +		+ +

+ = positive impact
+ /- = variable impact depending on pollutant and conditions

MACT (maximum available control technology) includes work practice or operational standards, including training or certification of operators to reduce the emissions of hazardous air pollutants.

This general training principle can be carried over into all operations in order to minimize or eliminate emissions. It is known that closing dampers or doors when equipment is on line has been sufficient to eliminate emissions by enhancing control features built into the equipment. This ensures that emissions are vented to a control device for capture and collection.

Regardless of the type of equipment, good housekeeping and maintenance practices are essential to ensure that emissions are kept to a minimum. Even the best state-of-the-art equipment still requires attention on a daily basis. Regular inspections and maintenance are required to ensure that the equipment has no leaks and is operating properly.

Some operation and maintenance practices include: proper waste handling, proper loading of equipment, maintaining adequate process times, proper operation of control equipment, and performing daily and weekly checks for leaks. Operation and maintenance practices have a large effect on the amount of raw materials used and the potential emissions of volatile and fugitive materials.

Fugitive Emissions

Other common operational problems are open inspection doors, poorly sealed duct work, and failure of operators to close all shutters. These lead to *fugitive emissions*.

Control of fugitives is receiving increased scrutiny and management attention. Fugitive volatile emissions refer to the minute amounts of process gases or fluids, typically organic, that escape to the atmosphere by way of a number of different mechanical routes. These include flange joints, sight glasses, packing and seals, valve stems and control valves, tanks and storage vessels, hose connections, unions, couplings, pumps, doors and gaskets. A typical valve leak at the stem packing is illustrated in Figure 6.1. Fine particulate matter is also of concern as a fugitive emission from granular solid material transfer points and storage piles.

Monitoring and replacement of leaking valves and flanges is a substantive part of good management strategies and housekeeping. In some cases, tightening of the packing nuts or bolts on flanges may be sufficient to reduce fugitive emissions. In other cases, it may be required to replace older valves with hermetically sealed equipment. Maintenance should not be forgotten, since failure to keep filters, fans, and air nozzles clean will upset the internal

Fugitive VOCs

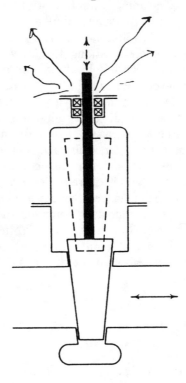

Figure 6.1.   Valve stem leaks.

air patterns of a dryer or other equipment and may result in "puffing" of contaminant-containing air from product entrance slots, instead of achieving overall air in-flow.

Daily inspections for leaks save raw materials and reduce fugitive emissions. Leaks can be in liquid or vapor form, continuous or periodic, depending on where the leak occurs. Liquid leaks are detected visually and are relatively easy to spot. Vapor leaks can be detected by a soap and water solution that is sprayed on the locations where leads may occur, or with more sophisticated equipment such as portable volatile organic compound analyzers. Performing daily leak inspections can prevent substantial fugitive emission losses through leaks that go undetected for several days.

Fluid pumps and packing around drive shafts in pumps may be sources of fugitive organic emissions. Monitoring, repair, and replacement of such pumps with magnetically-driven seal-less pumps are possible. This latter type of equipment (which has been determined to be BACT for refineries in Southern California) operates by having the pump impeller inside of housing which is sealed from the environment and in line with the moving fluid.

The driver transfers energy to the impeller shaft of the sealed unit by a magnetic field. This field, created by an outer magnetic ring, passes through a metal containment shell at the rear of the pump and turns an internal rotor, which drives the impeller, to produce the pumping action. The impeller shaft is supported by bushings lubricated by a stream of the process fluid. The mechanical seals, which are prone to leakage with externally-mounted systems, are thereby eliminated. Such a piece of hardware is typically designed for, and operates in, refineries and chemical plants where hydrocarbon-based liquids are being pumped.

Another technique to reduce fugitive emissions from open tanks and, in particular, plating solutions, is the use of additive chemicals. These surfactants reduce misting by lowering the surface tension of tank solutions. The addition of tank covers or floating surface media, similar to ping-pong balls, has been found effective in reducing surface evaporation and mists.

Product Storage Control

Breathing losses (vapor escape due to air and gas temperature changes from night to day) and working losses (vapor escape during filling) of volatile liquid organic compounds from storage tanks are a source of fugitive emissions. Control of emissions from product storage is performed by the use of one or more of the following techniques:

- floating roofs
- closed systems
- secondary systems

Emissions that escape from floating roofs and closed systems may be further controlled with secondary systems that recover the volatile compounds. This is accomplished by vapor recovery through condensers or absorbers. Floating roofs literally lie on the surface of stored liquids in a tank, thus preventing vapor generation. The only losses are at the seals between the tank wall and the floating roof edge. Figure 6.2 illustrates fugitive losses from the seals on a floating roof tank.

An important aspect of control of organic compounds is proper storage to prevent emergency or overfilling releases. Proper storage control can

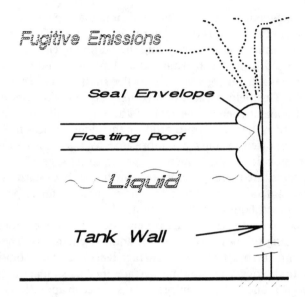

Figure 6.2.   Floating roof emissions.

prevent accidental releases. Monitoring is important to ensure proper product storage. Among the items that should be considered for monitoring are:

- fluid temperature
- liquid level
- liquid flow rate
- pressure
- tank stress
- spill containment capacity

Gases are stored under pressure, so two critical items to monitor are temperature and pressure. Liquids are normally stored at atmospheric pressure or a slight positive pressure, so the critical items to monitor for them are temperature and level. Liquid transfer, especially during loading and unloading, can be a point where spills occur either from accidental releases or through overfills. For these operations, it is important to check the equipment, the fluid flow rate, and the tank levels

Operator training and certification are additional management controls appropriate to these operations.

Materials Changes

Other management options include raw materials or compositional changes for the processes in question. These include activities which *minimize* the use of these raw materials, including greater efficiency of operation. Where possible, *substitutions* of materials which may be less volatile or possess inherently lower toxicity or are less reactive, may be considered. A common management approach is to perform surveys which identify those mechanical components which have high leak potential and institute management plans to ensure that leaks are (1) detected and (2) corrected with a minimum amount of leakage time.

It is also possible to change the composition of raw materials so that air emissions which are of a lower toxicity or volatility may be substituted in the process while maintaining product quality. One recent example of such a materials change was the *substitution* of a citrate cleaning solution for a chlorinated solvent used to decrease certain parts in the aerospace industry. The citrate compound, being a water-based system, virtually eliminated halogenated organic solvent emissions from that process line, which both saved money and eliminated hazardous air pollutant emissions.

Another example of substitution is in tank electroplating processes. Normally, aircraft parts are dipped in large vats of cleaning, rinsing, and plating solutions which contain cadmium to impart a special coating to the part. The use of aluminum ion vapor deposition as a substitute for cadmium tank electroplating was the focus of a large research project sponsored by the Air Force Civil Engineering Support Agency at Tyndall Air Force Base in Florida.

For those aircraft parts which do not have internal cavities or small clearances, it is possible to suspend the parts on a rack in a vacuum tank, which becomes the ion vapor processing chamber. The racks become the electrode of a high voltage circuit which, when the chamber is evacuated, gives the parts a negative charge.

A small amount of argon gas is bled into the system, and high voltage is applied to the parts. Thin strands of aluminum wire on the lower portion of the chamber are vaporized and, as they pass through the glow discharge set up by the ionized argon gases, the aluminum is ionized. These positively charged aluminum ions are then attracted to and are neutralized on the negatively-charged parts, thereby forming a corrosion resistant metal plating.

The ion vapor deposition process takes longer and costs more; however, the hazards from cadmium emissions coming from the tank plating operation are virtually eliminated. At the same time, contaminated waste waters from the electroplating systems are eliminated and likewise the hazardous sludge from electroplating tanks, which has to be treated as a hazardous waste under RCRA, is totally eliminated. In addition, there are no worker health and safety concerns, since the ion vapor deposition process is a totally enclosed system and no hazardous metals are utilized.

For large surface-coating operations, probably the most effective VOC reduction technique is one in which the paints and topcoats are reformulated to eliminate solvents entirely. In these processes, the resins, pigments, and other coating modifiers are applied in a powder state, again by electrostatic processes, and heated. This causes the resins to flow together, forming the smooth coat normally associated with solvent-based processes. Following the heating operation, the coated parts, such as automobile bodies, are run through radiant lamp arrays which cause the semiliquid material to cure into a hard and tough coating.

Switching from solvent-based to water-based paints and surface coatings is another technique which may be used to eliminate hydrocarbon emissions. Reformulation of the coating materials is required in order to disperse the pigment particles and resins evenly throughout the coating. This requires some additives, detergents, and/or small amounts of alcohols or ketones to decrease the surface tension of the fine particle sizes in the new coating. To some degree, these coatings are still experimental but are gaining acceptance in a large number of surface coating applications.

PROCESS OPTIMIZING ACTIONS

The second approach is to optimize processes for air pollutant emission reductions. A wide range of process changes are possible which lead to lower air contaminant emissions. One of the early attempts to modify processes was to "de-rate" the unit. This occurs by operating a process or system at less than its maximum capacity, which generally leads to a lower emission of air contaminants. This, however, leads to poor energy efficiency and lower production. It is, therefore, not high on the list of potential management options.

Many times it is possible to evaluate air emissions utilizing field testing equipment, whereby changes in the operation of a process are instituted in order to determine emissions as a function of those parameters. Many processes are sensitive to temperatures or speed of operation or other control room settings. With portable analyzers, it is possible to run a series of parametric tests of operational parameters versus air contaminant emission rates and then select those operating conditions or ranges where emissions are at their minimum.

Modifications of line speed, as well as processing time, may have significant bearing on air contaminant emissions. In one situation, emissions of CO and HF from a semiconductor plasma etch system used in the production of printed wiring board substrates were found to be proportional to the amount of time that it was operated. The process was normally run on a 30-minute mode which ensured that the product met the quality standards of the printed boards. It was found that shortening the time from 30 minutes to 22 minutes caused approximately a 40% reduction in CO and HF emissions with no change in product quality. Figure 6.3 illustrates this effect. The residual gases were then vented to control devices.

Another example is seen in gravure presses which coat inks onto continuously fed fiberboard or other substrate material. Emissions of VOCs from the evaporation of solvents in the inks occur as the web fed press moves the substrate material from one station to another prior to a dryer.

Operational improvements include reducing web speeds which will reduce the amount of solvent vapor "drag-out." High web speeds entrain air which draws additional solvent vapor from the ink coating operation into the open spaces between rollers.

Other common sources of evaporative solvent emissions would be insufficient temperature, ink that is too thick, slower drying ink than that for which the unit was designed, and under-designed dryer systems. Each of these problems may be addressed in a management or process optimization procedure which may reduce solvent losses by as much as half of the total solvent utilized.

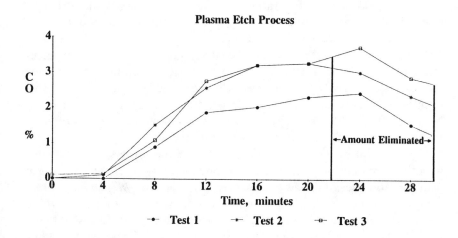

Figure 6.3.   Carbon monoxide emissions vs. time.

COMBUSTION MODIFICATIONS

Combustion modifications represent another area of contaminant reduction possibilities. Combustion, which contributes to NO_x, CO, and organic emissions, may be optimized (emissions minimized) by modifying how the combustion takes place. For instance, it has been found that lowering the usual amount of excess air in the exhaust gas yields lower oxides of nitrogen concentrations from boilers.

Oxides of nitrogen are key contaminants which form photochemical ozone and are largely emitted from combustion sources. A number of possible modifications may be made to the combustion process to lower NO_x without adding back-end hardware or changing the fuel.

Since NO_x formation is a function of the adiabatic flame temperature (Figure 6.4) as well as the stoichiometric air/fuel ratio, we find that most NO_x strategies attempt to reduce the maximum flame temperature as well as the residence time at these high temperatures. This "thermal NO_x" accounts for the majority of the NO_x formed in combustion systems and is larger than the NO_x from fuel nitrogen.

The first approach to lowering NO_x emissions in large multiburner boilers is to utilize what is called a "burners out of service" (BOOS) combination. Figure 6.5 illustrates such a BOOS combination which reduced the NO_x as noted. In this approach, a series of burners are taken out of service, while the balance of the fuel flow is provided to the *in service* burners.

Increasing the fuel flow through the burners which are in service decreases the local air-to-fuel ratio. The rest of the required combustion air is provided by leaving open the ports on the burners which are out of service. Thus, the overall NO_x is reduced.

Burner modifications are possible, contributing to different flame geometries and turbulence. Research is ongoing to modify burners to allow for *localized gas recirculation* and *delayed mixing* between fuel and air. This lowers the local air-to-fuel ratio and the flame temperature which reduces NO_x formation. Burners and air registers may all be modified to produce lower NO_x without changing the boiler or the fuel.

Tight control of the oxygen available to combustion system is possible with single or multiple burner external combustion systems. These "oxygen trim" or "low excess air" systems generally require *continuous monitoring* for exhaust gas oxygen, CO or combustibles, and oxides of nitrogen. If the monitor readout is in the control room, the operator is able to ensure that a stable and safe combustion process is always present. These systems allow for operation very close to the stoichiometric air-to-fuel ratio; however,

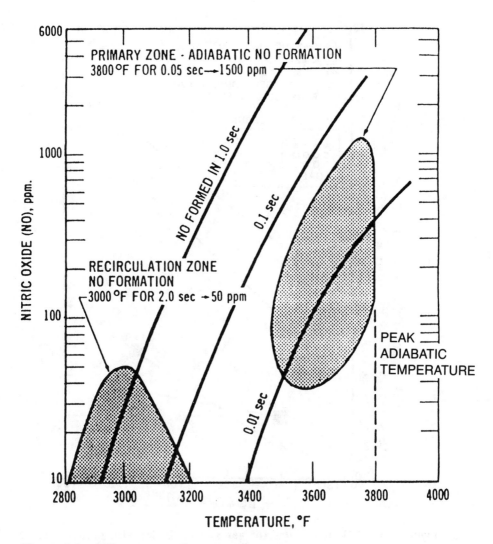

Figure 6.4.   NO$_x$ concentrations vs. temperature.

close attention must be paid to the level of oxygen in the overall system as well as CO and combustible gas levels.

It is possible to reduce the oxygen to too low a level, at which point the flame may become unstable. If so, it may begin pulsating at a natural harmonic frequency in the 10 to 20 Hertz range. At that point, the boiler

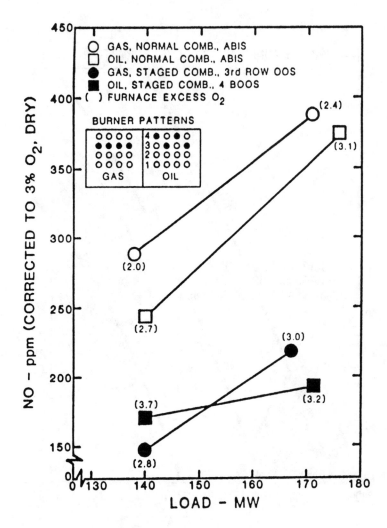

Figure 6.5. NO$_x$ reductions for gas and oil boiler with burners out of service.

itself may begin resonating in phase which, if unattended, will destroy the structural integrity of the unit.

Another option is to introduce water or steam into certain combustion processes, which will act to lower the flame temperature. This option finds its best utilization in combustion turbine emission reduction programs. Significant NO$_x$ reductions may be achieved by injecting liquid water or steam into the combustor can during operation. Figure 6.6 illustrates the NO_x reductions possible by injecting water into a 25 megawatt (MW) gas turbine driving a power generator. In this figure, we see the effects on NO_x for both Jet A and methanol fuels. Methanol without water injection presents a significant reduction by virtue of its combustion characteristics and lower adiabatic flame temperature.

For multiple burner boilers, it is also possible to provide a *staged combus-*

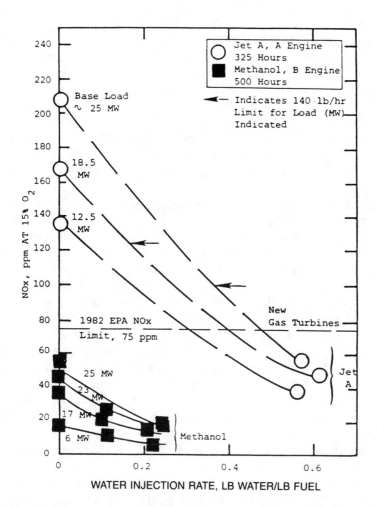

Figure 6.6.   Combustion turbine NO_x vs. water injection — liquid fuels.

tion in the furnace. In this approach different horizontal rows of burners are operated in either a fuel-rich or fuel-lean mode. In the fuel-rich zone near the burner NO_x concentrations are reduced. When the flames move up to the higher regions of the boiler where more air is available, the overall combustion becomes air-rich.

In the higher regions of the boiler, there are open air injection ports called *NO_x ports* which provide that final amount of air to complete the combustion and lower overall CO and combustibles concentrations in the flue gases. Oxides of nitrogen are reduced while the overall combustion process maintains its required heat output. Figure 6.7 is a schematic illustration of a power boiler with staged combustion zones.

Re-burning, often used in combination fuels, produces a very similar pattern, as seen in the previous figure. Re-burning involves passing lower burner fuel-rich combustion gas products up to a secondary flame. This process is designed to reduce NO_x without generating CO. It diverts a fraction of the total required fuel from the primary combustion zone burners to upper burners to create a secondary fuel-rich flame zone. Sufficient air is then supplied higher in the boiler to complete the oxidation process. Lab tests indicate that a maximum reduction in NO_x is achieved when the lower re-burning zone stoichiometry is approximately 0.9 (90% of theoretical air).

Flue gas recirculation is another potential process used to reduce NO_x by reducing flame temperatures. This requires a hardware modification and additional power, since up to 20% of the exhaust gas is recirculated back into the combustion air supplied to the burner system. In this approach, increasing amounts of flue gas are recirculated into the combustor, which lowers the amount of available oxygen. This acts to lower the overall combustion-produced thermal NO_x. Initially, a very large reduction for a very small amount of recirculated gas may be found (i.e., 50% NO_x reduction for a 5% gas recirculation).

With flue gas recirculation there are also concerns for space utilization, as well as capital and energy costs. This approach finds its greatest utilization in smaller package boilers, typically those operated with only one burner firing on natural gas.

It is possible to purchase low emission burners which give equivalent combustion temperatures and lower pollutant emissions. These burners, coupled with stack oxygen and CO monitors, allow significant optimization of combustion gas emissions.

Fuel modifications are another method of reducing air contaminant emissions. Fuels, being easier to change than hardware, are therefore an option for source reduction approaches.

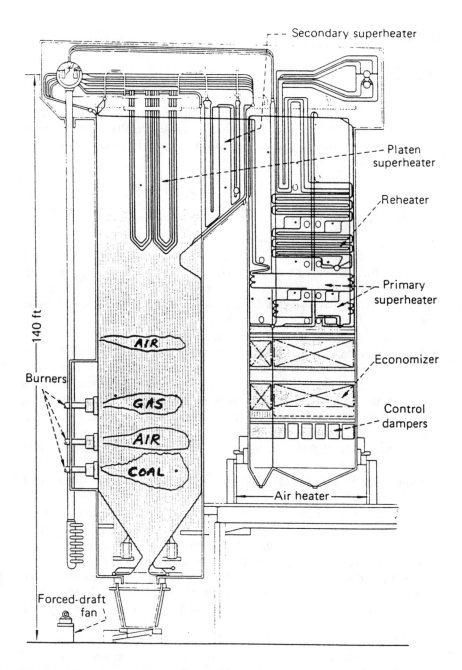

Figure 6.7. Power boiler with staged combustion.

FUELS

Efficiency

With respect to fuels, we may either minimize their use, modify them, or choose other alternatives. Processes which increase overall efficiency or can take advantage of lower steam production yield lower fuel use and therefore lower air pollutant emissions.

Secondary Utilization

Any operational or hardware change which minimizes or reduces the amount of fuel consumed *overall* in a combustion system will emit lower air pollutant mass emissions due to a lesser amounts of fuel being burned. Consequently, a great deal of attention has been focused recently on waste heat boilers, steam generators, and drying operations.

In these approaches, a high temperature utilization of heat, such as a boiler, incinerator, or other fired process, exhausts waste heat in the form of hot gases. These gases may be channeled into another process which is able to utilize the lower gas temperatures in a secondary process. Typical of such an operation would be a system in which combustion gases, after being used to make steam, are used to generate hot water or some other secondary use. The secondary use could be for building heat, to drive another process, or, in the case of natural gas-fired systems, to perform a drying operation as in the production of paper goods.

Some newer approaches use the *heat of condensation* of the water vapor in stack gases. This heat may be used for space heat or to provide a working fluid for another process such as a boiler or air conditioner. In these approaches, the stack gases which contain large quantities of water vapor are ducted through heat exchangers which extract the heat and lower the gas temperature to the water dew point. At that point, the latent heat of vaporization is given up in the heat exchanger, and the water condenses out to the liquid state. This provides large amounts of energy for use in other processes without requiring any further utilization of fuels or generation of air pollution.

Problems with this approach may occur due to the possibility of corrosive gases being present in the exhaust and the lack of buoyancy of the final stack gases. Contaminants, such as NO_2 or SO_2, form nitric or sulfuric acid in the water condensing in the heat exchangers, causing both corrosion and waste water disposal problems. Lower exhaust gas plume buoyancy could cause high noncondensible air pollutant concentrations downwind from a cooled exhaust due to lessened dispersion.

Fuel Switching

The simplest approach to lower emissions from fuels is to change the fuel itself. In general, coal is a "dirtier" (greater air pollutant emissions per unit of energy) but cheaper fuel than oil, and oil is a "dirtier" but cheaper fuel than natural gas. Therefore, where possible, switching from a solid or liquid fuel to natural gas could be a preferable alternative to building or adding an air pollution control system.

However, there are other concerns in fuel switching since combustors are designed for one type of fuel with a backup fuel as a temporary replacement. The distinction is in the size, temperature, and luminosity of the flame itself for a given fuel. In a large power boiler, changes in flame geometry due to a fuel change may have a significant impact on the temperature distribution in the hot water and steam passages in the boiler.

Fuel Blending

It is also possible to blend fuels for lower overall emissions. One such blend would be to utilize a pulverized coal boiler in conjunction with natural gas burners. This would lower the emissions of particulate matter and sulfur oxides and allow utilization of the cheaper fuel.

Apart from gas and coal combustion systems, the general approach is to blend high-sulfur and low-sulfur coals to meet emission limitations. Such a blending operation, particularly with a solid fuel such as coal to meet one emission limitation (SO_2), requires very tight control on the blending processes in conjunction with monitoring data from continuous emissions monitors.

In one such case in Detroit, a power plant was able to blend three coal types, low-sulfur western and southern coals with mid-sulfur eastern coals in varying percentages. The purpose was to balance sulfur content and cost and to meet the statutory SO_2 emission limit. In general, the low-sulfur western coals, which accounted for almost 50% of the blend, were required to meet the emission limitations. In general there are minor capital expenditures for such blended systems, as western coals are more prone to spontaneous combustion than the eastern coals. Likewise, additional costs for reducing boiler fouling and slagging are usually incurred.

Depending upon the location, a switch from a high-sulfur coal to a low-sulfur coal is possible, again depending upon the SO_2 limits. However, a complete switch from one to the other can have far-reaching implications for both operation and emissions.

Coal varies considerably in moisture, ash, sulfur, and heating values. As seen in Chapter 3, many low-sulfur coals have a lower heating value than the higher sulfur coals; therefore, more fuel is needed to produce the same amount of energy. Therefore the amount of ash and moisture that will pass through the system to maintain the power output will require additional efforts to maintain boiler efficiency. Disposal of waste products, such as fly ash and bottom ash which may be two to three times higher requires additional space and cost for disposal of residuals.

Fuel Cleaning

Coals are particularly amenable to cleaning since the major contaminants are commingled minerals and soils. Favored techniques to beneficiate coal are by grinding the coal, washing it with water to remove the coal particles (which tend to float) from the rock and soil fractions. These latter fractions are much denser than coal and fall to the bottom of the water wash. Other possibilities include *froth floatation*, in which air is also injected into the water-washing step. In this process small bubbles adhere to the coal particles which help them rise to the top of the water. At that point, the coal is discharged from the top of the fluid, dried, and used as fuel. The solid soils, minerals, and rocky material are separated and disposed of as refuse. Figure 6.8 illustrates one type of coal cleaning unit.

It is possible to grind coal and separate it by an air classification process. In this process pulverized coal is discharged over an air curtain. The coal particles, being less dense, float on the air cushion or jet until they are collected and sent to bunkers for storage. The more dense materials tend to fall out of the air cushion and are removed as solids. In modifications to this process, large cyclones (q.v.) are used to separate out *pyrites* which are much more dense than the coal particles. Pyrites are iron sulfide minerals, so their removal reduces the sulfur oxide and ash/particulate emissions. This approach avoids the problems of dealing with water solutions and coal drying before utilization. It is not as effective in removing other noncombustible materials. It is also subject to fugitive particulate losses.

For fuel oils, it is possible to clean them of water soluble contaminants. In one situation in California, certain fuel oils contained significant

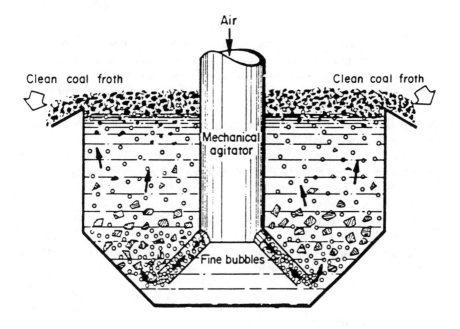

Figure 6.8.   Froth flotation cell.

amounts of salt. This contributed to boiler fouling, slagging, increased corrosion, and increased emissions of sulfuric acid droplets, rusty particulate matter, and acid fallout in nearby communities. In one field demonstration it was demonstrated that by injecting fresh water into the fuel oil during transfer the dissolved salts were transferred to the water solution. Following a period of roughly two days, the water droplets containing the salt coalesced by gravity separation into the bottom of the storage tank. This reduced the oil's total salt content from over 100 ppm to less than 15 pm. This greatly beneficiated the fuel oil and cut the slagging, fouling, and air contaminant emissions dramatically. The contaminated water was removed by a simple decanting operation which went to a water treatment operation.

Natural gases may be beneficiated by control of sulfur content. From an air pollution perspective little else is necessary to beneficiate gas, due to the small amounts of contaminants naturally occurring in gas.

Additives

Fuel oil additives are sometimes used in an attempt to lower air emissions and to control fouling of boiler surface. They influence the size of fuel oil droplets in combustion systems by changing the viscosity and surface tension. Changing the droplet size will change the evaporation rate, which will change the heat flux. Combustion parameters and particulate emissions are influenced by these changes. Some reductions in visible plume opacity have been noted in past trials with additives.

Fouling increases emissions indirectly by lowering thermal efficiency (which requires greater fuel). Fuel oil additives influence fouling by modifying the chemical composition of tube deposits inside the boiler. The primary effect is to produce a high melting point ash deposit that is powdery or friable and easily removed by a soot blower or a lance for control by back end control systems. When the ash is dry, corrosion is reduced, and large particulate emissions may be reduced. In some cases, alkaline materials are added in attempts to reduce acid gas emissions.

Fuel Modifications

Natural gas, oil, and coal will form the backbone of fossil fuels for the foreseeable future. However, modifications of some of these fuels, in particular those involving coal, will find their place as "niche" markets.

Mixtures of crushed coal and oil in slurries are one attempt to utilize a resource in plentiful supply. The benefit of such a coal-oil slurry is that a relatively cheap fuel may be utilized with the advantages of oil. The primary advantage of liquid fuels is that they are easily pumped, and combustion parameters and fuel mixing are more easily controlled. Also, particulate emissions and sulfur oxides are lower than those of pure coal combustion.

Other very specific blends of fuels include adding oxygenated compounds, such as alcohols or ethers, to gasoline to improve the emissions characteristics of the fuel. The effect of adding oxygenated components to liquid fuels has the effect of "leaning out" the air-to-fuel ratio, which results in reduced hydrocarbon and CO emissions. The addition of oxygenated fuel components serves to reduce flame temperatures (and NO_x) and to reduce the emission of heavier molecular weight compounds that are more difficult to oxidize. The net effect of such reformulated gasolines is to increase the combustion efficiency and lower air emissions.

The other modification is to take a primary fuel, such as coal, and convert it to a synthesis gas. In this approach, a fuel which is basically carbon, like coal, is reacted at high temperatures in the presence of steam to yield a mixture of carbon monoxide and hydrogen gas (Equation 6.1):

$$\text{Carbon (C)} + H_2O \rightarrow CO + H_2 \qquad (6.1)$$

This synthesis gas may then be used as any other gaseous fuel without the problems of ashes, particulate matter, or pyritic sulfur entering into the combustion reactions. This "water-gas reaction" is beneficial; however, it requires the input of heat and high pressure and leaves behind a solid waste residue which will need disposal.

Pyrolysis of solid fuels is another approach in which a synthesis gas may be generated from a lower-grade fuel. In this process, the solid fuels (of whatever type) are heated to drive off moisture and then they are raised to higher temperatures to drive off the volatile organic materials. These gases may then be used directly in a combustion process or piped elsewhere for utilization. In general, these approaches for pyrolysis fuels are limited by the distance between the pyrolyzation process and the location of the usage of the gas.

Depending on the availability of other components, such as hydrogen or natural gas, the fixed carbon may be further oxidized to a CO gaseous fuel which likewise may be used as a synthesis gas. These pyrolysis-derived gaseous fuels are generally utilized internal to a facility where the raw materials are readily available and a large quantity of synthesis gas may be used in close proximity. These onsite uses are to provide steam to a process or hot exhaust gases to dry out other materials or provide heat for other plant reactions.

The alternatives are the *biogenic* fuels which are generated by the operation of natural processes (i.e., fermentation) on biomass, such as wood chips, walnut shells, etc.

Methyl alcohol and ethyl alcohol are two alternative fuels which may be generated from the biological degradation of wood, grain, and biomass to yield an alcoholic short-chain hydrocarbon. After separation, these alcohol fuels may be used directly or as an additive to a composite-fuel mixture. The problems associated with these fuels are the amount of heat and processing required to convert biomass materials into a cleaned, dried alcohol. Of course, residual materials must also be disposed of as solid waste.

Biomass materials may also be used directly as solid fuels. Wood chips and walnut shells have been used in place of other fuels in pulverized combustion systems and in hogged fuel or waste fuel boilers. Where a supply of these materials is plentiful and the transportation distance is short, such fuels may find a niche in providing process heat or steam. However, these materials tend to be lower in heating value than the primary fossil fuels and generate particulate emissions. The latter erode gas passages, ductwork, combustors, and process equipment.

Other biomass fuels, such as landfill gas, may be used either directly onsite to provide energy to a combustion turbine or a reciprocating engine to directly drive a power generator. They may be *beneficiated* by removal of moisture, carbon dioxide, and other heavier molecular weight constituents and then sold as an upgraded "natural" gas or fired in a conventional boiler as a natural gas substitute. Again, the cost implications are that, due to the lower heat content of landfill gas, it is generally more cost-effective to burn it directly and provide either steam or power onsite.

Fuel Refining

For liquid fuels, it is possible, through additional refining or reworking of the basic fuel hydrocarbons, to lower the sulfur content and residual minerals to provide a beneficiated fuel. This provides a cleaner fuel for the end user but leaves the residuals materials sulfur, ash, etc., at the refinery for disposal.

Reformulation of fuels and specification changes to lower the amount of heavy molecular weight compounds may result in cleaner combustion, fewer deposits, better efficiency, and lower emissions. One such measurement is the boiling point of 90% of the liquids in the fuel. This is represented by the upper right-hand portions of the distillation curves seen earlier.

A recent study found that specifying a lower T_{90} (the boiling point of the 90th percentile of the weight of fuel) resulted in lower emissions of nonmethane hydrocarbons, benzene, formaldehyde, acetaldehyde, and 1–3, butadiene with no changes in CO emissions and a small increase in NO_x emissions. This study, called the "Auto/Oil Program," provided some significant information on compositional changes for gasoline emissions.

PLANNING AND DESIGN

Primarily for new sources, air quality management options have their biggest impact at the design stage rather than as a back-end control system or a source modification. In general, these approaches involve the entire stationary source from location to layout to types of technologies in use.

Geographic Location

The first effort would be to choose a location, where possible, which optimizes ventilation and natural dispersion. A location that is more open to horizonal wind movements and unlimited vertical mixing would be preferable to one in which a low inversion or even a narrow valley location is available. In this scenario, the emitted gases that may be lost to the atmosphere are more easily diluted by natural processes. This is assuming that process changes, material substitutes, etc., have already been implemented in the planning stages to reduce potential emissions before the facility has been designed.

Where possible, the location may be chosen as a function of zoning. Placement of a facility in an industrial location, or in one with a sizeable distance between industrial or commercial districts and residential areas works to abate nuisance complaints due to odors and particulate fallout. In addition, "hot spots" (localized areas with high concentrations due to point sources) may be minimized.

In planning a layout of a facility, the minimization of fuel, raw materials, or end product movement will inherently lower potential air contaminant emissions. Minimizing fuel movement, either by tanker, truck, or by pipeline, will lower fugitive losses as well as catastrophic spills or leaks.

Minimizing handling and movement of raw materials, providing storage *enclosures* rather than open storage, and enclosing conveyor belts, storage bins, etc., will work to lower the amount of fugitive emissions. Minimizing movement of products will generally lower overall fuel requirements for forklifts or for electrical power consumption in moving materials or products from one location to another or to the final loading dock.

Low-Emission Systems

In a newly designed facility it is possible to take advantage of new technologies which will give greater overall efficiency of energy use and thereby lower costs and fuel combustion emissions. For steam or power production, such operations could entail combined cycle facilities. In one such case, a combustion turbine, firing on natural gas, provided 15 megawatts of electrical power which was then sold to the local utility, after which the hot exhaust gases were used to dry paper goods without deleterious effects.

Other combustion technologies, such as fluidized beds, may be used where gaseous or even solid fuels may be burned to provide controlled temperatures. In a fluidized bed combustor the fuels are dispersed in a swirling mass of solid material, such as sand, and ignited. The temperature of the hot gases is carefully controlled to provide a uniform temperature. These temperatures are relatively low, but the sand or other media has high turbulence which contributes to complete combustion with uniform heat transfer. Fluidized bed systems can be used to carefully control emissions in an NO_x-sensitive area while providing high-quality steam for other processes. Dirtier fuels, such as coal, may also be utilized in such a system, since it is possible to add alkaline materials, such as limestone, which will react with acid gases during combustion.

Where possible, new energy technologies may be utilized to provide local power or heat which require less intensive conditions. These include the potentials for fuel cells and solar systems to provide energy for specific processes. These are, of course, limited by the relatively low energy density of such systems and find only a limited use in specific niches.

In designing processes or buildings, "energy-efficient" facilities are a contributing factor to lower emissions, since less energy has to be used to provide space heating or cooling for the occupants.

EMISSIONS CHARACTERIZATION

After all source reduction, management, and planning activities have been carried out, it is necessary to fully characterize air pollutant emissions from a given process prior to the design or installation of back-end controls. Air pollution control systems or *back-end* techniques will always be required, because there will always be some emissions from any process, as no process is 100% efficient in the conversion of raw materials or fuels to end products.

The first step in control system selection is to characterize the gas stream *which will be carrying the contaminants* from the source to the control system. Velocity, total gas flow rate, water content, temperature and pressure of the exhaust gases are critical. Each of these may significantly influence the control device, as well as the materials used to contain the flow and collect the air contaminants. It is *not* sufficient to know just the average values. It is important that one also understand the maximum possible range, both on the high and the low side, for each parameter, since any control system must handle excursions or process upsets.

Likewise, an understanding of the emitting process itself is needed. This includes the mass throughput, the temperature, reaction rates and ratios over the course of the process. Likewise, one must fully understand whether the process is a *steady-state* one (coating a material on a web or substrate) or whether the process is a *cyclic* one (i.e., batch chemical manufacture) with surges both in temperature, mass, and volume throughput and/or emission rates. Control room parameters which determine process rates must also be monitored, since they are generally indicative, directly or indirectly, of air contaminant emissions.

Emission changes are illustrated by the production of steel in a basic oxygen furnace (BOF). Here pure oxygen is injected, by a water-cooled probe, into molten pig iron in the BOF. The oxygen rate is modulated to provide a burn-off of excess carbon in the bath until the appropriate level of carbon (and steel composition) is reached. This process produces highly variable emissions of particulate matter and CO, and illustrates that any control system must provide for such surges or variations.

For each contaminant one must understand its physical state (gas, solid, or liquid) and its concentration. Both extreme values and average concentrations are important. Likewise, the mass emission rate of such contaminants must be fully characterized, lest the control system be oversized, causing capital cost concerns, or undersized, with the potential for emissions of undesirable air contaminants.

The chemical and physical properties of each contaminant must also be fully characterized. Chemical properties of the contaminants include knowledge of whether the material is organic or inorganic and its reactivity to other gases in the gas stream, to moisture or to oxygen. These may significantly impact the choice of a control system. Likewise, the oxidation potential of the contaminant, as well as its pH and corrosivity in either the gaseous or liquid state must be known. When exhaust gases pass over a "cold spot" which induces moisture condensation, contaminants may dissolve in the liquid and concentrate in those areas of the control system. Such a point might be an uninsulated area in the duct work, an expansion joint, a heat exchanger, etc.

The physical properties of the contaminants are of concern as well. The water solubility of each contaminant in the gas stream will affect the choice of control system, as well as the potential discharges of waste water generated in the system (either intentionally or unintentionally). If the contaminant is a particle, its size, density, and "abrasiveness" of the particulate must be known. Knowledge of whether the particulate may be "sticky" can affect the choice of control system and the duct work design. The ducting containerizes and conveys the exhaust gases and particulate matter from the process source to the control device.

COLLECTION OF AIR CONTAMINANTS

If a collection system is not provided as an integral part of the source equipment, such as duct work and chimney for a boiler, provisions must be made for collecting the contaminants. These include direct emissions (i.e., those escaping from a metallurgical operation) as well as fugitive emission occurring from transfer points or open surface operations.

Knowledge of whether the contaminant is a vapor or a particulate is required in designing or evaluating a contaminant-capture system. Both types of contaminant emissions are momentum-controlled; however, particles are less influenced by gas movement than vapors.

Provisions for containerizing the emission by a hood or a shroud during normal operations, movement of raw materials and final products, and maintenance are important. The temperature and buoyancy of the gases escaping from a surface, such as an open top degreaser or a molten metal bath, must be known, since they will significantly affect the direction of movement of the contaminants into the surrounding air. Figure 6.9 illustrates a typical hooding arrangement to collect emissions from an open-top process, such as a metallurgical operation.

Once the contaminant gases are containerized, they must be conveyed by a local exhaust system to the control device. Particulates are of the greatest concern since their density and inertia influence whether they will "fall out" (by sedimentation) in the exhaust ducting prior to the control device. The

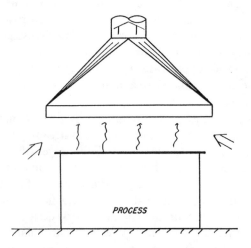

Figure 6.9.   Canopy hood over a process tank.

conveying velocity needed to suspend particles is a direct function of the particle shape and density and may vary from as low as 2000 feet per minute to as high as 5000 feet per minute. The shape of the ducting is another important consideration in order to avoid recirculation zones or dead zones which will cause particles to drop out and clog the system.

One of the key parameters to optimize in an air pollution control system is the total amount of air that the air pollution control system must handle. In a fugitive control system, such as an open hood, a large amount of air must be drawn into the system in order to capture the particulates and gases that may be emitted. However, the intake velocity to capture particulates decreases exponentially with distance from the ductwork entrance. Therefore, very large quantities of air must be drawn in to provide sufficient velocity at points farthest from the hood to guarantee contaminant capture.

These concerns have led to innovative ways to utilize other air properties to lower the amount of total air drawn into the system. Such a system would be one in which a "slot hood" may be placed along the sides of a liquid-containing tank. The tank may contain a water bath, acid bath, molten metal or solvent. Slot hood designs take advantage of *air* entrainment over an open surface. The higher the velocity, the greater the entrainment. Thus, a long slot with a very thin opening just above the surface of a molten or liquid bath would entrain more capturing air at the far side of the tank than an equivalent contaminant capture system with an overhead or suspended hood. Figure 6.10 illustrates such a slot hood design for control of fugitive emissions from a liquid bath.

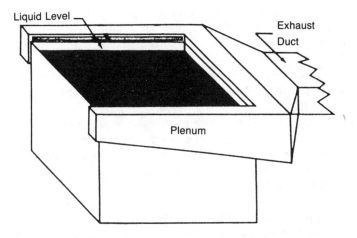

Figure 6.10.   Slot hood for control of emissions from open-surface tanks.

Additional refinements to the slot hood approach have been to provide a "push-pull" system which takes advantage of additional entrainment of air in a high-velocity jet. These small compressed air jets sweep across the surface of the liquid bath aimed right at the suction of the slot hood. This provides for even greater contaminant collection at a fairly low flow rate. A push-pull system requires a source of compressed air, but this is considered a nominal cost due to the lowered expense of smaller fans and control systems.

Other concerns for exhaust systems and control devices are for the materials of construction. Key concerns are for *abrasion or erosion* of the gas passage materials when particulates are present, as well as concerns for fracture or breakage of control or exhaust system components under thermal stress. This occurs when there are rapid temperature swings in the process. Leaks may result.

Corrosion may also be induced by the formation or use of liquid water in the process. The combined effects of moisture, temperature, and air (quite apart from corrosive gases) may cause weak points or areas under thermal stress or thin wall areas to rust. These areas may suffer abrasion or be corroded through, causing losses of collection efficiency and increased emissions.

Of concern from an economic standpoint is the sizing of the equipment to allow for proper gas flows and controls at an optimum cost. In general, the horsepower of the exhaust fan is one of the largest single determinants of cost. An understanding of fan laws, as they relate to energy consumption, is important to factor into exhaust control system designs. The fan laws are:

- Fan volume varies directly with speed
- Fan pressure varies with the square of volume
- Fan power varies with the cube of the volume

The total volume for venting any given process is a direct function of the speed of the fan. If we double the fan speed, we double the volume of gas being drawn through the control system.

The total pressure of the system increases as a square function of the volumetric flow rate. Thus, if the volume is doubled, the total pressure required to be exerted in by that fan is four times the original amount. Thus a significant pressure increase requirement results from doubling the volume of gas being exhausted.

Of particular concern, since electricity costs are based upon the amount of power utilized, is the third fan law. This says that if we double the volume of gas in the same system, the power requirement will be eight times higher, since the power increases with the cube of the volume being drawn through the system. Therefore, the costs associated with a control system are significantly influenced by the required power. This requires that the volumetric flow rate be carefully selected.

Even an increase in volume of 25% would lead to nearly a doubling of the power requirements and costs associated with it.

AIR POLLUTION CONTROL APPROACHES

For convenience, "back-end" control systems are divided into those designed for gaseous emissions and those designed for particulate matter. It should be remembered that few sources emit only one contaminant; therefore, most air pollution control systems must take into account the entire range of contaminants, temperatures, concentrations, and energy costs. Likewise, control systems designed for one pollutant often may have some effectiveness for other ones.

Multiple control devices in series may be utilized for contaminant control. In these systems, a lower efficiency simpler device is usually placed first to collect the bulk of the contaminant emissions, while the second device functions as a "polishing" unit.

Gas Control Technologies

With gases, control systems are divided into two categories: *process gases* and *combustion gases*. The process gases include *volatilized organic compounds* such as solvents (i.e., perchloroethylene), fuel vapors (i.e., propane, hexane), organic chemicals (i.e., toluene, methanol, etc.) and *acid gases*. The acid gases include hydrochloric acid, nitric acid, sulfuric acid, phosphoric acid, SO_2, and NO_2. The *combustion gases* are carbon monoxide, products of incomplete combustion (fuel fragments), sulfur oxides, and NO_x.

For each of these gases only a few types of control equipment are generally available. The type of equipment usually describes the physical or chemical process that occurs in the equipment. Which one, or combination of devices, may be chosen depends on the gas stream and pollutant characteristics noted earlier. It is therefore more appropriate to survey the equipment types and their general attributes.

For VOCs, for instance, the *concentration* of the organic compound strongly influences the selection of a control device. Where the concentration is in the percent by volume range (i.e., 30 to 40%) as in the displaced vapors from filling an underground gasoline storage tank, a condensing process using a chiller is the most useful system since it recovers the vapors for reuse.

Where the VOC concentration is in the low parts per million range (i.e., from solvent evaporation), an adsorption system such as activated carbon is useful because the adsorber acts at the molecular level to collect the molecules of solvent with high efficiency. Also, with that system the solvent may again be recycled. Of course other VOCs, regardless of concentration, must be incinerated due to either regulatory requirements or their high toxicity potential. Figure 6.11 illustrates typical ranges for VOC control technologies as a function of their concentration.

Absorbers

Absorption is a gas-liquid contacting process that utilizes the preferential solubility of the pollutant gas for the liquid phase. Contact between the gas and liquid phase is by mechanical means, and this is done by providing as much of a transfer zone as possible. This transfer zone is usually provided by a packed tower, a spray chamber, or a plate scrubber, with the gas phase usually passing countercurrent to the liquid phase direction.

Typical applications are for gas streams containing organic compounds including alcohols, acids, substituted aromatics, aldehydes, and esters. The following are certain points to consider for these applications:

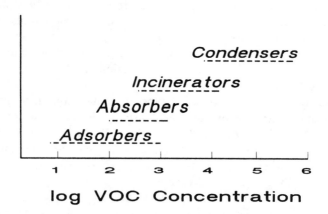

Figure 6.11.   Technologies for VOC control.

- The absorption process is highly temperature dependent – Henry's law (relative solubilities of organics in the gas phase as opposed to the liquid phase) applies.
- Water treatment may be needed for the water solutions to remove the stripped material or chemical reaction products.
- Some organic compounds, while dissolved originally in the water, may outgas in a conventional waste water treatment system.
- Acid gases are collected by using water or alkali-containing water solutions to absorb and neutralize the acid.

Figure 6.12 illustrates the typical absorption tower.

Adsorbers

Adsorbers work on the principle of molecular or atomic interaction at the surface of a sorbent. There is a decrease in free energy of the system; therefore, the process is always exothermic regardless of the contaminants involved. The usual application is with gas phase air pollutants.

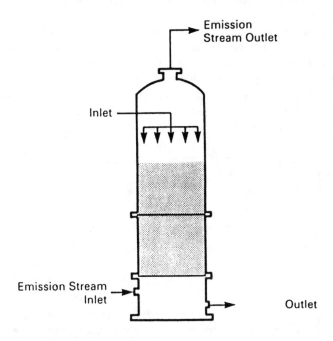

Figure 6.12.   Packed tower absorber.

Adsorbents are composed of either natural or synthetic materials with a micro structure whose internal pore surfaces are accessible to the gas phase. Some typical adsorbent materials are: activated carbon, activated alumina, molecular sieves (zeolites), and silica gels.

The adsorbent is usually regenerable and may be used many times in the control process. This suggests the use of at least two units in parallel on a continuous gas stream so that fresh adsorbent is always available. One unit is shut off from the gas stream and regenerated; the other unit is on line, adsorbing pollutants in the gas stream. A variety of regenerant gases (typically steam or dry nitrogen) are used, depending on the pollutant, the adsorbent, and the availability of steam and/or water treatment facilities.

Typical applications include gas streams containing volatile organic compounds with molecular weights over 45 but not greater than 130 (if over 130, they are difficult to desorb). The adsorption technique has application to some gaseous toxic materials, principally organic chemical compounds. Compounds principally adsorbed are chlorinated organics, alcohols, ketones, and aromatics. To be considered for this application are the following items:

- Whether the collected pollutants desorb easily or are retained on the adsorbent.
- If there is a mixture of compounds being collected from a gas stream, regeneration is sometimes difficult or impossible. In this case the adsorbent must be treated offsite, potentially as a hazardous waste, if the compounds are classified as hazardous.
- Waste water treatment may be needed for the organic compounds that are absorbed during the regeneration process; steam, when used for regeneration, condenses out in a liquid water phase along with the desorbed organics.
- Some organic compounds that are absorbed in the regenerant condensate and sent to treatment may volatilize and become an air pollutant in conventional waste water treatment systems.

An activated carbon adsorption system is illustrated in Figure 6.13.

Condensers

Condensers operate by the removal of heat from the gas stream, and provide a surface or medium for condensation to take place. There are two types of condensers: surface and contact. In surface condensers, the coolant

STEAM REGENERABLE ACTIVATED CARBON SYSTEM DESIGN

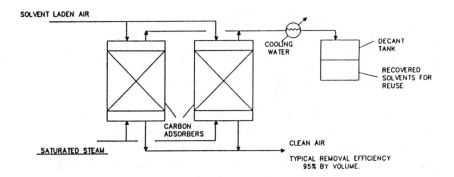

Figure 6.13.   Activated carbon adsorber.

does not come in contact with the vapors or liquid condensate. In contact condensers, the coolant, vapors and condensate are all intimately mixed.

Surface condensers are usually of the shell and tube or fin type. The coolant usually flows through tubes in the shell and tube types, with the condensate forming on the outside. Some air-cooled condensers pass the condensing gas stream through the tubes and cool air over the tubes. With the fin-type condenser, vapors condense inside the tubes while cooled by a curtain of falling water. The tubes may be completely submerged in water.

Contact condensers operate by the liquid coolant coming into direct contact with the condensing vapors. These are usually simple spray chambers with baffles to provide adequate contact. Others are high-velocity jets (ejectors) to produce a vacuum. The contact condenser generally removes more pollutants than the surface type, but it may result in a waste stream that requires treatment.

Condensers are often used in combination with other control devices to remove some or all of the condensable vapors such as organic compounds. Typical applications include gas streams containing aldehydes, alcohols, chlorinated compounds, fuel vapors, and organic acids. There are certain points to consider:

- The process is energy intensive.
- Corrosion may occur on the surface of the condenser.
- Waste water treatment may be necessary if a contact condenser is employed.

Thermal Oxidation

Many organic compounds released from manufacturing processes can be controlled by rapid oxidative combustion. Oxidation systems include thermal oxidation units (or incinerators), catalytic combustion units, and flares. Rich or high concentration organic gas streams are also used in boilers for fuel or as a supplemental fuel.

One type is *direct flame incineration* where a combustion chamber is employed. The direct flame unit is used where the gas stream is below the lower flammability limit (LFL) and therefore cannot sustain combustion. The direct flame incinerator turbulently mixes the gas stream with air in a fixed combustion zone and is fired. The retention time is usually 0.5 to 0.75 sec. The mixing and retention time allows combustion of the weak gas stream.

Incineration is an excellent destruction method for most organic pollutants in a gas stream. Some inorganic and metallic compounds should not be sent to an incinerator. The major disadvantage of incineration is the cost of supplemental fuel needed to burn most of the compounds in a high temperature combustor.

Catalytic systems operate by passing the premixed gas stream over a heated bed surface coated with noble metal catalysts (usually containing platinum or palladium). The catalytic surfaces allow the organics to be oxidized at lower temperatures than incinerators with equivalent destruction efficiency.

Typical applications include gas streams containing aldehydes, esters, reduced nitrogen compounds, saturated and unsaturated alkyl halides (but not with catalytic oxidation), carboxylic acids, olefins, and aromatics.

For this application one must consider:

- If the incinerator is a catalytic type, one must be aware of any catalyst "poisons" such as phosphorous, or heavy metals in the gas stream.
- Gas streams containing sulfur or chlorine are usually subject to additional controls.
- The residence time and temperature necessary to completely burn any toxic compounds.
- The cost of fuel and equipment as compared to other control alternatives.

PARTICULATE TECHNOLOGIES

Particulate matter is controlled by a series of different control technologies. These include mechanical collectors, electrostatic precipitators, baghouses, wet scrubbers, and combination units such as venturi scrubbers. In general, the particulate removal is a function of the particle size. Figure 6.14 illustrates the typical collection efficiency of a variety of particulate control devices as a function of the particulate size.

Of particular concern for these control devices are their relative efficiencies for trace element removal. Health and environmental effects from trace element emissions may significantly impact the health risks of various control technologies. Table 6.2 indicates the removal efficiencies of various control technologies for a variety of trace elements in the exhaust gas from conventional stationary combustion processes. As a practical matter, the higher the energy requirement to operate these devices, the more effective the control technology is for collecting and removing trace elements.

Mechanical Collectors

All mechanical collectors operate by gravity or inertial forces such that the mass and aerodynamic particle size are used to separate the particle from the gas stream.

The types of mechanical collectors are:

- settling chambers
- cyclones
- dry inertial-type collectors

Settling chambers depend on the inertia of the particle so that it will drop out of the gas stream due to gravity when the gas stream is passed into an expansion section, with a resulting decrease in gas velocity. Simple settling chambers are not efficient collectors for particulate diameters less than about 40 to 50 microns. Another problem is that if gas velocities increase above 10 feet per second, the particles may reenter the exhaust.

Cyclone separators are the most widely used type of particulate matter collection equipment. The particulate matter follows the gas stream along the wall of the cyclone and down to a collection chamber. At the vortex of the cyclone, the gas stream changes direction and goes up and out of the cyclone, leaving the particulate matter to fall out of the gas stream and into

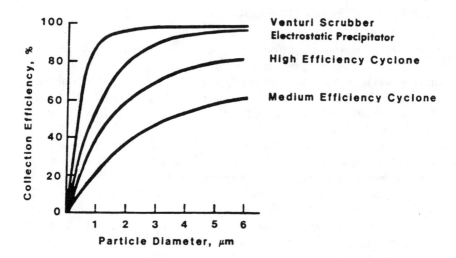

Figure 6.14. Fractional collection efficiency of commonly used dust collectors.

Table 6.2. Typical Particulate Matter Trace Element Removal Efficiencies of Various Control Technologies

	Average Trace Elements Removal Efficiencies (%)					
				SO$_2$ Scrubber[a]		
Trace Elements	Cyclone Separator[a]	Electrostatic Precipitator	Venturi Scrubber	Coal-fired Boiler	Oil-fired Boiler	Baghouse
Aluminum	66.0	99.2	99.6	99.0	92.0	~100
Arsenic	75.3	95.3	94.2	97.0	81.0	ND
Beryllium	84.3	98.4	99.2	98.0	ND	ND
Cadmium	44.0	95.6	92.3	99.0	77.0	ND
Chromium	27.7	95.1	92.5	95.0	90.0	ND
Iron	54.2	99.1	>99.5[a]	99.0	95.0	~99.9[a]
Lead	30.0	95.5	98.0[a]	99.0	94.0	~100
Mercury	3.2	0.0[a]	12.6[a]	55.0	87.0	ND
Nickel	18.6	52.5	95.0	95.0	83.0	~100
Selenium	33.1	86.0	91.4	87.0	97.0	ND
Titanium	74.4	98.9	99.8	ND	ND	~100
Zinc	39.4	97.0	98.4	98.0	90.0	~100

[a]Does not represent an average value, since only one data point was available.
ND – No data available.

the collection hopper. Cyclones are generally not efficient at collecting particulate matter less than about 5 microns. A cyclone mechanical dust collector is seen in Figure 6.15.

Other types of dry inertial-type collectors include centrifugal collectors with a fan device, louver-type dust separators, and baffle chambers. None of these types of mechanical collectors are very efficient for particulate matter less than about 20 microns in diameter.

Mechanical collectors are not useful as the principal collector for particulates containing toxic materials. They do serve well as a precleaning device in series with other, more efficient, particulate control devices by reducing the particulate loading in the gas stream. Pressure drops range up to several inches of water column.

Fabric Filters

Baghouses or fabric filters have a wide application for the control of particulate matter. The fabric provides the support for the filtering mecha-

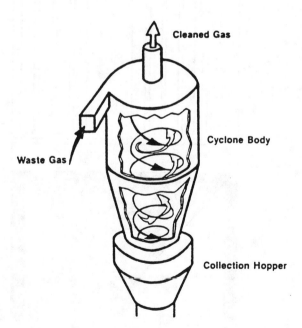

Figure 6.15.   Common cyclone dust collector.

nism, a thin layer of particles (either from the gas stream or induced) known as the *pre-coat*. It is the pre-coat that forms a *filter cake* on the surface of the fabric. Figures 6.16 and 6.17 illustrate a typical baghouse and a cut section of the filter cake on the bag fabric during operation.

The filter fabric selection depends on the gas stream conditions and cleaning mechanism. Gas stream conditions to consider are:

- temperature
- particle size
- abrasiveness of the dust
- gas face velocity (through the bag)

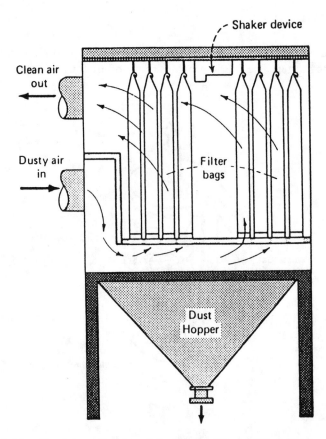

Figure 6.16.   Single-compartment baghouse filter with shaker.

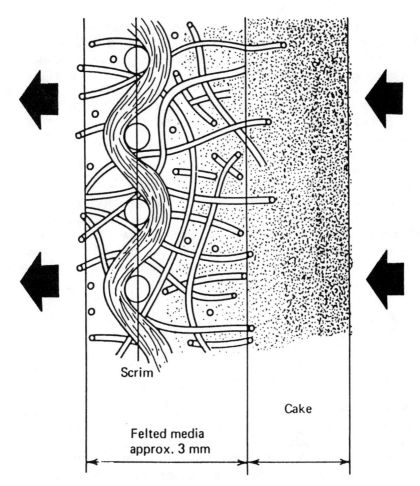

Figure 6.17.   Felted media showing distribution of dust.

Fabric filters are good collectors for particulate matter if the gas stream is not too hot (generally below 450°F for fiberglass bags is the upper limit) and the particulate matter is not wet. Often, baghouses can be used in combination with other control devices. Pressure drops range between three and six inches of water column (in. w.c.). Space requirements in the plant may be of concern if a retrofit application is considered.

Typical applications are for controlling particulate matter emissions such as dusts, soils, metals, and dry solid organic compounds. They have been installed for years on sources as large as coal-fired electric power plants. In many cases, alkaline dust is used as the pre-coat to aid in SO_2 removal. Baghouses are not effective where hot gas excursions are anticipated, since the bag fabric determines the unit's operating temperature. Also, they are not effective when condensable particulates may form. This is significant because some metals and hazardous air pollutants may form a condensable particulate fume when the gas stream is cooled to less than the upper gas temperature limit.

Wet Scrubbers

Wet scrubbers provide control for a wide range of particulate emissions. Collection is by impaction with, or diffusion of the particulate matter to, the water droplets. When they are used to control gaseous emissions, it is through the process of chemical absorption (see absorbers). There are a number of different types of wet scrubbers. Some of the more common types for control of particulates are:

- spray chambers
- plate (tray) systems
- centrifugal
- dynamic (wet fan)
- venturi

Scrubbers used to control dry particulates in gas streams are usually equipped with a water spray system. In addition to large particles, the sprays quench high temperature gas streams and condense out some organics and inorganics. Typical applications include gas streams containing emissions of dusts, metals, "sticky" particulates, and soluble organic compounds.

Figure 6.18 illustrates a venturi scrubber system with a cutaway section showing the "throat" area where the gases and dust interact with the fine droplets. In addition, venturi systems have the capacity to handle hot gases, condensable compounds, a wide range of particulate chemical and physical properties, acid gases, and wet sticky or abrasive particles. A key advantage is that there are no "moving parts" in the unit. Space requirements may be minimal for vertical installations.

For high efficiency control of particulate emissions, venturi scrubbers are considered effective control devices. While other scrubbers do control particulate emissions, they are not as effective in the submicron particle diameter range as the venturi units.

Venturi scrubbers are very efficient particulate collection devices for par-

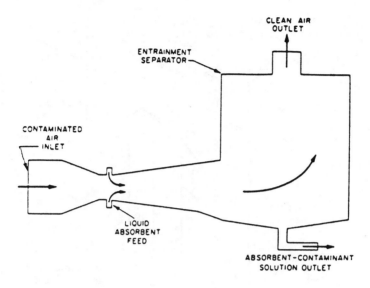

Figure 6.18.   Venturi scrubber.

ticle diameters greater than one micron. The efficiency in the submicron range can be increased by using a high energy venturi scrubber with a pressure drop across the venturi throat of 45 to 60 or more inches w.c. This is considered a significant energy penalty and presents a serious consideration for any installation.

Important points to consider for this application are:

• particle size
• condensable materials in the gas stream
• chemical and physical properties of the particles
• Waste water treatment for the scrubber liquor

Electrostatic Precipitators

Electrostatic precipitators (ESPs) are control devices for particulate matter with a specified range of electrical resistivity. They also have applications in controlling mists and condensable organics such as oils. The ESP operates on a principle of electrical ionization and charging of particles or droplets in the gas stream. The charged particles or droplets migrate to a collecting electrode as they pass through an electric field. A typical electrostatic precipitator section is shown in Figure 6.19.

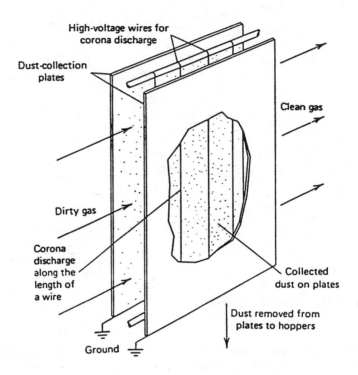

Figure 6.19. Diagrammatic sketch of a single-stage precipitator.

Dust is collected on a dry plate and is mechanically removed or "rapped." It then falls into a hopper. Mists and condensed oils or organics are usually collected in a "wet" ESP where the plate is either washed on-line with a water solution, or the unit is shut down and the plates cleaned mechanically.

Dry ESPs work well with most dry particulate-laden gas streams. They can be scaled to handle large gas volumes. They do not work as well as other methods for the collection of condensed particulate matter and do not work well with sticky particulate. When the particle diameters are less than about 0.5 microns, dry ESPs do not work as efficiently as other control technologies.

Wet ESPs have been applied to gas streams containing metal fume and sticky particulate and generally find more applications to toxic chemicals than do the dry ESPs. Gas streams such as acid mists, sticky organic material, and metal fumes are controlled better with the wet than with the dry ESPs. Typical applications include gas streams containing particulate emissions of metals and metallic compounds. The "wet" ESPs are also used to control "tars" such as emitted from pitch impregnation and asphalt felt saturation processes. Pressure drops are generally quite low, ranging down to a few tenths of an inch w.c.

There are certain points to consider for this application:

- The gas stream temperature: wet ESPs do not work well above about 170°F. Dry ESPs may suffer from corrosion if the gas stream temperature is too low.
- Water cleanup may be needed for wet ESP effluents.
- Particle resistivities must be in a limited range for efficient collection.
- Space requirements may be quite large.
- The particle size to be collected is generally greater than 0.5 microns.

COMBUSTION GAS TECHNOLOGIES

Combustion gas controls include equipment operation and back end control technologies that have been demonstrated to have a high or proven potential to reduce these emissions. They do not include the *combustion modifications* discussed earlier.

Carbon Monoxide and Combustible Gases

Carbon monoxide and combustible gases may be destroyed by either high temperature incineration or use of catalytic systems. Incinerators are the same as those used for organic gas control. In general, under normal excess oxygen combustion circumstances, elevating the temperature of the exhaust gas to above 1400°F with a residence time in excess of 0.3 seconds is sufficient to destroy CO and combustibles. Operating boiler or combustion systems in the overall oxidative (or lean regions) of the combustion zone is sufficient to control these oxidizable carbon gases.

Sulfur Dioxide

Sulfur dioxide is considered a combustion-generated gas. Control techniques have been utilized where alkaline materials, either liquid sprays or solid dusts, have been injected into gas streams containing SO_2. These alkaline materials accomplish neutralization of sulfur dioxide within the alkaline droplets or on the injected particles. These neutralized particles and/or liquid droplets are then collected by other control devices placed later in the system.

Oxides of Nitrogen

Oxides of nitrogen present a significant challenge since they are present in low to moderate concentrations (i.e., between about 50 to 250$^+$ ppm) in hot gases, usually at high volumetric rates. The up-front or combustion modification systems are used in the largest number of applications for cost reasons, but back-end control approaches are rapidly gaining favor, particularly where emission limitations are tight. There are primarily two processes for NO_x control: the *reduced nitrogen* and the *oxidative* processes. The reduced nitrogen processes are better known by the two approaches — *selective catalytic* and *selective noncatalytic* based systems.

The reduced nitrogen processes attempt to chemically react with the NO and NO_2 molecules and reduce them back to nitrogen gas and water vapor. These take the form of post-combustion controls by either directly injecting ammonia, or by supplying a catalyst in conjunction with it such that Equation 6.2 occurs:

$$NO_x + NH_3 \rightarrow \rightarrow N_2 + H_2O \qquad (6.2)$$

Injection of liquid sprays or dusts containing reduced nitrogen compounds with the $\cdot NH_2$ functional group, such as urea, has been found to be effective in reducing NO_x by up to 70%:

$$NO_x + \cdot NH_2 \rightarrow \rightarrow N_2 + H_2O \qquad (6.3)$$

Table 6.3 summarizes the potential emission reductions of SO_2 and NO_x for a variety of combustion gas controls in coal-fired utility boilers.

Selective Catalytic Reduction (SCR)

Selective catalytic reduction reduces NO_x emissions using a catalyst bed and ammonia. Ammonia, taken from a storage tank, is vaporized and diluted with air or steam to produce gaseous ammonia. The ammonia is injected upstream of the catalyst and mixed with the flue gas. The flue gas along with the ammonia then enters a catalyst reactor, which contains several layers of catalyst elements (a catalyst bed) and is distributed throughout

Table 6.3. NO$_x$ and SO$_2$ Stationary Source Control Technologies

Technology	Potential Emission Reductions (%)	
	SO$_2$	NO$_x$
Commercial		
Fuel Switching and Blending	50–80	0–10
Physical Coal Cleaning	20–50	0
Low NO$_x$ Burners	0	30–50
Overfire Air	0	15–30
Lime/Limestone FGD[a]	90–95	0
Dual Alkali FGD	90–95	0
Spray Drying	70–90	0
Near Commercial		
Integrated Gasification Combined Cycle	90–95	90–95
Fluidized Bed Combustion	80–90	>50
Selective Catalytic Reduction	0	80–90
Furnace Sorbent Injection	50–70	0
Low-Temperature Sorbent Injection	50–70	0
Reburning	15–20	35–50
Emerging		
Advanced Coal Cleaning	45–60	0
Electron Beam Irradiation	80–95	55–90
Copper Oxide FGD	90–95	90–95

[a]Flue Gas Desulfurization.

the catalyst bed. The catalyst enhances the reaction of ammonia with NO$_x$. When the proper conditions are obtained, such as a flue gas temperature between 275°C and 424°C, a reaction sequence takes place and NO$_x$ is formed into molecular nitrogen and water, thereby reducing emissions.

Increasing the ammonia injection rate leads to an increase in NO$_x$ reduction. However, this can result in an increase of unreacted ammonia (ammonia "slip") which can cause blockage of gas passages, poor emissions control downstream, and visible ammonia salt plumes. Ammonia slip can be minimized by matching the NO$_x$ concentrations profile in the flue gases at the inlet of the catalyst. Also, an ammonia control system can be used to monitor the ammonia dosage applied to the SCR system.

The catalyst configuration is usually of a honeycomb type with parallel rigid plates and cylindrical pellets. Typical catalyst materials are titanium and zeolite. The lifetime of the catalyst is limited due to losses from poisoning, erosion, solid deposition, and sintering by high temperatures. Once the catalyst loses its activity, it must be replaced. The lifetime of a catalyst is important in terms of cost and maintenance.

SCR is applicable to gas, oil, and coal fired combustors, and results in NO_x reductions of 80% to 90%. Unfortunately, the cost for SCR installation and maintenance is high.

Selective Noncatalytic Reduction (SNCR)

When the reduced nitrogen/NO_x reaction takes place entirely in the exhaust gases without the presence of a catalyst, the process is called selective noncatalytic reduction. Three common chemicals (in effect, additives) used with SNCR are ammonia, urea, and cyanuric acid, each with a different effective reaction temperature range. The latter two may be in either a liquid solution or in a powder/dust form. The additive is injected at various distances downstream of the combustion chamber of the boiler, heater, or other combustor. The objective of SNCR is to inject the chemical or solution at an optimum *temperature window* (between 900°C and 1100°C), where the reaction will take place to convert NO_x into nitrogen and water.

If, however, the exhaust gas temperature is too high, ammonia will react with oxygen to form more NO_x. If the temperature is too low, the ammonia does not fully react and results in poor NO_x reduction and excessive ammonia slip. Good mixing of ammonia and flue gas NO_x is essential to achieve a high NO_x reduction and low ammonia slip.

The solid or liquid spray additives are used to expand the *temperature window* to lower temperatures. Urea injection has the advantage of not being hazardous, as is the case with ammonia or cyanuric acid.

SNCR equipment consists basically of sprays, temperature sensors, a chemical feed control system, additive storage tank, and a chemical vaporizer. Reduction in NO_x emissions up to 75% can be achieved with SNCR. Selective noncatalytic reduction has the advantage of not requiring the expensive catalyst equipment, but does require a higher consumption of chemicals.

Oxidative Systems

Oxidative systems also find their niche in NO_x control. In these systems, which are typically specific to smaller gas flows with relatively high concentrations of oxides of nitrogen, the exhaust gases are processed through an oxidative absorption system. Here, intimate contact occurs between the oxides of nitrogen and solution chemicals in a packed tower. In solution, the oxides of nitrogen are fully oxidized to nitric acid and absorbed. The resultant solutions are neutralized in a wastewater treatment process.

Table 6.4. Relative Effectiveness of Stationary Control Technologies[a]

System	VOCs	HAPs	PM	NO$_x$	CO	SO$_2$
Absorbers	+ +[b]	+ +[c]	+			+ +
Adsorption	+ + +	+ +				
Condensers	+ +	+				
Cyclones			+			
Fabric Filters		+	+ + +			+
Oxidative						
Catalytic	+ +	+ /−		+	+ /−	
Thermal	+ +	+		+ /−	+ +	
Reductive				+ + +		
Electrostatic			+ +			
Scrubbers			+ +	+		+ + +

[a]Effectiveness varies over a wide range, depending on system parameters. For general comparisons only.
[b]Oxygenated organics.
[c]Acid gases.

In these systems there is a potential safety problem in handling materials such as ammonia and in disposing of residual materials such as wastewaters following wet control systems. Back-end systems for NO$_x$ control may be used *in conjunction with* combustion modifications and low nitrogen fuels for more complete NO$_x$ reductions.

TECHNOLOGY COMPARISONS

A comparison of the applicability of various control systems is seen in Table 6.4. This compares the relative applicability of individual control options for the range of criteria and noncriteria contaminants seen in stationary source emissions. As can be seen, no one technique is uniquely qualified to be a "cure-all" for all types of air contaminant emissions. Rather, a comprehensive approach, taking into account contaminant-specific parameters, location parameters, efficiency, and economy are all required to produce the most effective air pollution control system.

It should be remembered that regulations may become increasingly stringent. Therefore, new facilities must put a premium upon flexibility of control systems, incorporating possibilities such as changing fuels, tighter emission controls, and higher operating or maintenance costs. Concerns for residuals waste management must also be addressed. Thus, the emphasis in the future will increasingly be on front-end management, planning and source reduction, in addition to higher efficiency back-end control technology.

Control System Hardware Considerations

A final note should be made regarding the selection of hardware and the key factors involved in that hardware. Probably the first two major concerns are the efficiency of the control system to meet the technology or health-based emission standards and the cost which allows the system to be built.

As noted earlier, control efficiency requirements can be expected to increase, since tighter emission limitations will be adopted in the years ahead. Therefore, systems designed to attain higher than originally required efficiencies of collection and control will be considered better options. Detailed economic studies comparing the various control efficiencies as a function of capital and operating costs, with estimates of potential break-even points, are considered in all capital purchases. Increasing concerns are now expected for the "life cycle cost" of any technology. With the increasing emphasis on environmental protection of all media, not just air pollution, the costs of residuals management must also be factored into cost analyses.

The durability of any given system to withstand the temperature, process and concentration excursions with changing mixtures of contaminants is also a factor to be considered. The more durable the system, the less maintenance time and cost would be required to keep the system operational.

The reliability of any system is high on the list of hardware considerations. It has been noted that anything will work for a while, but when large capital outlays and high pollutant control efficiencies are required, estimates of reliability are a major consideration. Not to be forgotten are the potentials for *liability* (both corporate and individual), should equipment fail to perform as required or, worse, a release of hazardous air pollutants occurs.

In general, simplicity of operation and maintenance are rated higher than a nominally equivalent or slightly higher control efficiency of another more complicated system. Potential problems with component part breakdowns which may cause releases above emission limits are also to be considered. Matrix analyses for control systems, bringing into play technological, legal, and economic factors, will increasingly be a tool in control system design and selection for stationary source air quality management.

7   MOBILE SOURCE CONTROL APPROACHES

Anyone who thinks we should return to the horse and buggy should consider the emissions from a horse.

late night comedienne

Motor vehicles comprise the largest number of sources of air contaminant emissions in the United States. These include passenger vehicles, light duty trucks (LDTs), and medium to heavy duty trucks (M/HDTs). As seen in Table 7.1, passenger and light duty trucks form the majority (approximately 95% of the total) of the number of motor vehicles on the road.

The significance of these sources is that they are primarily gasoline-burning. All mobile sources in this table are internal combustion engines, but the specific type of combustion varies. Gasoline-powered vehicles are responsible for about 80% and 91%, respectively, of the gallons of fuel consumed and miles driven. Diesel fuel oil accounts for the balance. Indi-

Table 7.1.   U.S. On-Road Motor Vehicles — 1990

Vehicle Type	Number (millions)	Fuel Gallons (billions)	VMT (billions)
Passenger	126	74	1616
Light Duty Trucks	46	24	404
Medium/Heavy Duty Trucks	9	26	200
Total	181	124	2220

VMT = vehicle miles traveled.

vidual emissions from each source are small; however, due to the large number of sources, the aggregate emissions are significant, as seen in Chapter 4.

ENGINES AND AIR POLLUTANTS

On a pollutant-specific basis, mobile sources account for varying percentages of air contaminant emission. Figure 7.1 indicates that these vary from 70% of the total national CO emissions to less than 30% of the particulate matter emissions. However, on a specific geographic basis, such as in California, mobile source emissions may be significantly higher. The internal combustion engine is the basic power plant for these vehicles, whether spark-ignited or compression-ignited.

There are also a significant number of nonvehicular internal combustion engines which may play a significant part in air quality management strategies. It is estimated that between 7 and 8 million outboard engines are in use in the United States at the present time, and 8 to 12 million engines in lawn mowers, leaf blowers, chain saws, and similar applications. These emissions are basically uncontrolled, and therefore their total contribution, in addi-

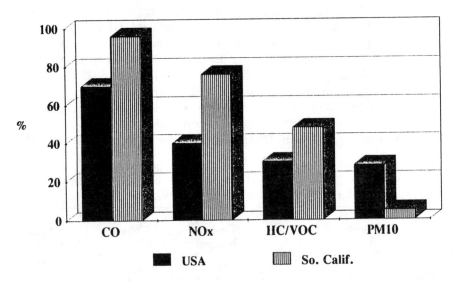

Figure 7.1. Mobile source emissions (contributions by percent). Source: 1991 National AQ and Emissions Trends & 1989 California ARB Inventory.

tion to those from aircraft, may represent a significant local impact on air quality. The principles involved in understanding air pollutant emissions from internal combustion engines (ICEs) are the same.

Likewise, stationary source IC engines operating on the same principles will have the same pollutant formation patterns; however, they are not subject to the changes in operating modes typical of a mobile internal combustion engine. Thus, stationary gasoline- or diesel-powered *reciprocating internal combustion engines* have the same emission patterns but are usually operated in a "cruise" mode rather than the cyclic pattern of mobile sources which change from idle to acceleration to cruise to deceleration to stop. Thus, emissions attributable to changes of operating mode are not significant for stationary ICEs.

The most significant difference between mobile sources driven by piston engines and those driven by combustion turbines is that the latter exhibit continuous combustion. The processes occurring in piston engines are essentially a series of explosions internal to the cylinder which show tremendous differences in temperature, pressure, gas composition, and volume occurring throughout the cycle.

One aspect of mobile source emissions is that they comprise both combustion emissions and evaporative or fugitive emissions. In addition, there are hydrocarbon emissions during refueling operations.

Engine Thermodynamic Cycles

The purpose of any mobile source of emissions is to provide useful work to drive a vehicle, whether automobile, truck, or airplane, to a different location. To accomplish this, useful work must be extracted from the engine.

Useful work out of a system is described by its thermodynamic cycle. The three cycles representing mobile sources are illustrated in Figures 7.2, 7.3, and 7.4 by their respective pressure-volume (P-V) diagrams. In piston engines, the cycles are represented by what happens to the fuel and air mixture in the cylinder. These four steps of the cycle are: compression, combustion, expansion or work-producing (step), and exhaust. These four steps are the same whether it is a two stroke or a four stroke engine. (The strokes refer to the number of times the piston traverses the length of the cylinder *for each power step.*)

Figure 7.2 refers to the spark ignition cycle, commonly called the Otto cycle after the individual who built the first successful operating spark-ignited gasoline engine. This figure illustrates each of those steps on an arbitrary pressure volume diagram. Beginning in the lower right-hand cor-

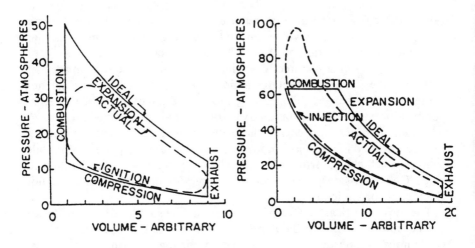

Figure 7.2.   Otto cycle.
Compression ratio
of 9:1.

Figure 7.3.   Diesel cycle.
Compression ratio
of 19:1.

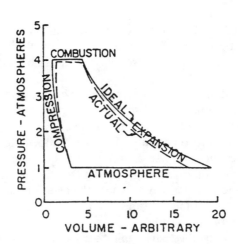

Figure 7.4.   Brayton cycle.
Pressure ratio 4:1.

ner, the fuel and air are compressed to much smaller volume with a slight increase in pressure, at which point ignition occurs. From that point, combustion begins, which dramatically increases the pressure in the system with only slight variations in volume. As the combustion proceeds, expansion of the hot combustion gases increases the volume and decreases the pressure as useful work is accomplished. The last step of the cycle is when the combustion gases are exhausted from the cylinder. This brings us to the point of return where the cylinder is ready for a fresh charge of fuel and air.

Figure 7.3 illustrates the diesel cycle in which much higher compression of the air in the cylinder occurs. At a point near the maximum pressure, fuel is injected (sprayed) into the cylinder near the final portion of the compression stroke, after which combustion occurs by auto ignition. The hot gases provide work to the drive shaft by expansion against the piston. This increases the volume and lowers the pressure in the cylinder until the gases are exhausted.

The thermodynamic cycle for combustion turbines is seen in Figure 7.4. This is the Brayton cycle, and operates at significantly lower pressures. In this system, compression is accomplished by compressor blades, and combustion occurs via a standing flame at nearly a constant pressure. Exiting the combustor can, the hot gases expand and drive turbine blades (which power the compressor) and exit the exhaust, providing thrust for the aircraft.

A summary of the significant differences in ignition source, pressure, air-to-fuel ratio, peak temperatures and peak pressure is seen in Table 7.2. These are useful in evaluating performance-based emission differences later.

Table 7.2.   IC Engine Operating Parameter Comparisons

	Otto Cycle[a]	Diesel Cycle[a]	Brayton Cycle[b]
Ignition source	Spark	Compression	Spark
Peak pressure, atm	35	100	4
Compression ratio	6 to 11:1	14 to 22.1	N.A.
Fuel delivery	Aspirated or injected	Injected	Injected
Air-to-fuel ratio	Near ideal	Lean	Very lean
Peak temperature (°R)	4500–5000	4500–5000	2250–2600

[a]Reciprocating piston.
[b]Continuous, combustion gas turbine.
N.A. = not applicable.

POLLUTANT FORMATION IN SPARK-IGNITED ENGINES

Formation of air pollutants in spark-ignited IC engines occurs in two regions: the bulk gas region and the boundary layer, or surface region. Each region has unique properties; therefore, the relative amounts of criteria pollutants and their formation mechanisms differ in each region. These regions are illustrated in Figure 7.5.

Bulk Gas Emissions

The bulk gas reactions for a spark-ignited engine include both fuel hydro-carbons and CO, and are generally formed by similar mechanisms. Oxides of nitrogen formation occur solely in bulk gas reactions, and are a function of many variables. Particulate matter is a significant contaminant for diesel engines and is addressed primarily through back end controls.

Hydrocarbons and CO tend to form by two mechanisms in the cylinder, depending on whether they are in the fuel-rich or the fuel-lean condition at any point in the thermodynamic cycle. In a fuel-rich condition, hydrocarbon fuel fragments and CO will be formed due to a deficiency of oxygen to support complete combustion. Such fuel-rich conditions occur during start-up, deceleration, and warm-up periods. In the locations where the gas mixture is in a fuel-lean (excess air) condition, oxidized carbon gases will be formed and remain, due to incomplete flame propagation. In these regions, carbon monoxide tends to predominate.

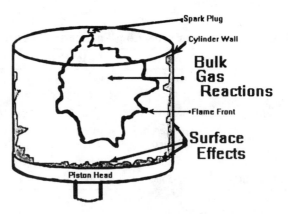

Figure 7.5.   Regions of IC engine pollutant formation.

Carbon monoxide once formed is thus fixed by the chemical kinetics of reactions in the bulk gases. Carbon monoxide is difficult to oxidize without high temperatures; therefore, its formation occurs as a result of *thermal quenching*. This effect is rapid at high air-to-fuel ratio mixtures. Hydrocarbon levels depend more strongly upon the amount of oxygen present. The effect of thermal quenching for hydrocarbons is much more severe for a given temperature gradient than it is for CO.

Oxides of nitrogen formation are a function of many variables, including the gas temperature (for thermal NO_x), the residence time at high temperature, and the availability of excess oxygen. The latter is a function of the air-to-fuel ratio.

The bulk of the oxides of nitrogen are formed in the hot, turbulent gas regions of the flame. The thermal NO_x formation mechanism is called the Zeldovich mechanism. In these high temperature regions, molecular oxygen is dissociated into oxygen free radicals which react very quickly with nitrogen to yield one NO molecule plus a nitrogen free radical. The nitrogen free radical then attacks an oxygen molecule to yield one NO plus an oxygen free radical, and so on. Equations 7.1, 7.2, and 7.3 illustrate these steps in the formation of NO by the Zeldovich mechanism.

$$O_2 \rightarrow 2O* \tag{7.1}$$

$$O* + N_2 \rightarrow NO + N* \tag{7.2}$$

$$N* + O_2 \rightarrow NO + O* \tag{7.3}$$

Thus, for every oxygen molecule which is cleaved by high temperature, four NO molecules will form while regenerating oxygen free radicals. There is therefore a near exponential increase of NO with temperature as the percentage of oxygen molecules being cleaved increases.

Surface Formation Region

The other major region of air pollutant formation is at the walls and surfaces in a cylinder. The cylinder of a spark-ignited IC engine serves as a very large *heat sink,* as well as providing high surface areas for physical and/or chemical reactions. Also, for gases within the cylinder, a boundary layer of fuel and air will form along the surfaces of the piston head and walls, which significantly influences emissions formation.

The walls and head of the cylinder and piston are a major source of hydrocarbons, carbon monoxide, aldehydes, and other products of incomplete combustion (PICs), due to the quenching of combustion resulting from heat sink temperature losses. It has been estimated that approximately 1% of the entire fuel charge is not burned as a result of these wall effects.

Deposits, as well as cracks and crevices in and on the surfaces of the cylinder, will enhance the trapping of fuel hydrocarbons in such deposits or crevices. Deposits are formed from localized hot spots, causing corrosion or localized cold spots. These carbonaceous deposits act like a fuel vapor "sponge." During the varying temperature regimes of the cycle, these sponges act to adsorb and desorb fuel components and products of incomplete combustion.

As the piston moves up and down in the cylinder, a film of oil forms on the walls, yielding a "wet" effect. This wetted wall serves as an additional location for absorption or desorption of fuel fragments. These factors contribute to hydrocarbon and PIC emissions during operation.

Four-Stroke Pollutant Mechanisms

An illustration of one cylinder typical of a gasoline powered IC engine during the four "strokes" of normal operation is seen in Figure 7.6.

During the compression stroke when the fuel and air are in the chamber, oil and deposit layers absorb hydrocarbons. Fuel and PICs (from the previous cycle) are forced into cracks and crevices in the cylinder surfaces.

During the combustion stroke, the pressure is still rising as the spark from the spark plug ignites the entire mixture. As the flame front moves through the mixture, NO forms in the high temperature burning gas. CO, if the mixture is fuel rich, will be present in the high temperature gases. Due to the increasing pressure at this point in the cycle, unburned fuel will be further forced into crevices on the surfaces of the piston head and exposed cylinder walls.

During the expansion stroke the piston is forced downward, and the volume begins increasing in the chamber: the temperature begins dropping, and therefore the NO formation is frozen as the burned gases cool. That is followed by a freezing of the CO combustion chemistry. Along the walls and from crevices in the cylinder, an outflow of hydrocarbon fuel fragments from those crevices begins. Some portions of those hydrocarbons will form CO and products of incomplete combustion.

During the exhaust portion of the cycle, the pressure in the cylinder drops to slightly above atmospheric, and wall effects begin to dominate. Deposits, cracks, and crevices desorb additional hydrocarbons, fuel fragments, and PICs. Desorption of fuel fragments from the oily layers along the walls of

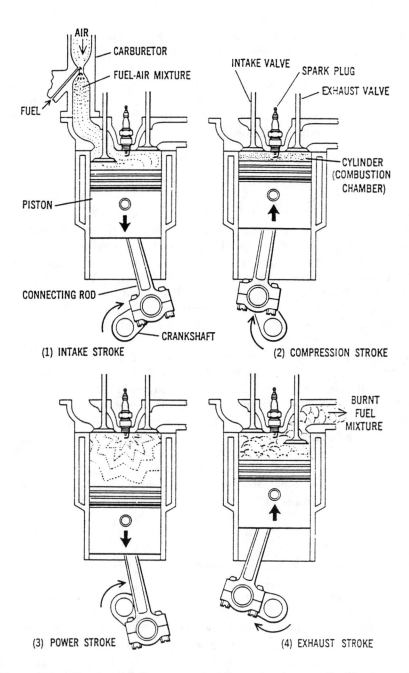

Figure 7.6.   Combustion in an automobile engine (one cyclinder of a typical automobile engine shown).

the cylinder will occur. The cylinder head begins scraping further fuel from the walls and desorbs those into the exhaust gases prior to the closure of the exhaust valve.

From this, we may understand some of the basics of air pollutant formation in an IC engine. These are a function of the complex interactions of pressure, temperature, volume, combustion kinetics, and mechanical effects in a spark-ignited gasoline-powered engine.

Lesser Sources of Carbon Gas Pollutant Emissions

The effects of wear and aging on engines may contribute significantly to hydrocarbon and CO emissions. These are partly due to the formation of surface deposits or corrosion building up over the course of time. Poorly seated valves and rings may also cause leaks of fuel or fuel fragments into the exhaust. Likewise, poor or faulty ignition generates pure hydrocarbon emissions during cranking. Likewise, scoring and crevice formation on aging engine surfaces lead to high emissions.

"Blowby," which is the flow of fuel past the cylinder walls into the crankcase may be a significant source of uncontrolled hydrocarbon emissions. In older uncontrolled vehicles, these may account for 20% to 25% of the total hydrocarbon emissions. Newer vehicles recirculate crankcase gases by positive ventilation systems back into the combustion air intake for reburning.

Scavenging losses will occur when both intake and exhaust valves are open at the same time. In a two-stroke engine, where both valves must be open for the engine to operate, scavenging losses are a major source of hydrocarbons, since seating of the valves and design of the combustion chamber require them to both be open during portions of the cycle. For a four-stroke engine, the scavenging losses occur when a supercharged or turbocharged system is in operation. This causes portions of the fuel-air mixture to pass directly from the intake to the exhaust.

Fuel Composition and Exhaust Emissions

Figure 7.7 illustrates the variety of air pollutants (organic compounds and fuel fragments) in the exhaust of a spark-ignited gasoline-powered four stroke engine. The four categories are the paraffins (saturated hydrocarbons), the aromatics or unsaturated ring structures, the olefins or double-bonded carbon systems, and the oxygenates or fuel fragments containing oxygen. These categories are charted by *species percent* for each carbon number represented. The actual number of individual compounds in gaso-

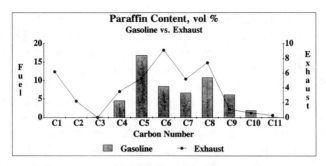

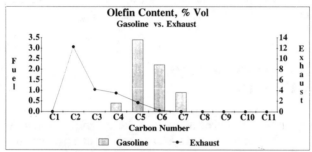

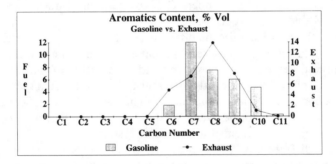

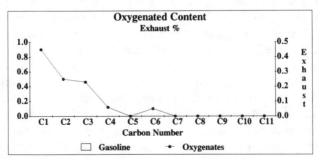

Figure 7.7.   Exhaust gas distribution vs. fuel composition.

line and its exhaust ranges into the hundreds of discrete chemical species. This figure also illustrates the typical average gasoline composition, listed by species. Oxygenates are the partially burned fragments of fuel left in the exhaust. Interestingly enough, the largest single oxygenate is the single C1 species *formaldehyde*. The single C2 compound is acetaldehyde.

A comparison of the fuel composition with exhaust hydrocarbon composition demonstrates the strong correlation between the exhaust species distribution and the fuel species. For paraffins and olefins, there is also a "downshift" to lower carbon numbers representing fuel fragments. This indicates that, except for the oxygenates, the exhaust hydrocarbon emissions are *components* from the original gasoline.

DIESEL IGNITION EMISSION CHARACTERISTICS

The significant differences between a diesel-ignited system and a spark-ignited system are that the diesel system operates at extremely high pressures (approximately 100 atmospheres) and high (lean) air-to-fuel ratios, producing high excess air in the chamber. The bulk gas temperature range, as seen earlier in Table 7.2, is about the same.

One of the more significant differences, though, is that diesel engines operate by injecting a measured amount of *oil* into the cylinder at high compression. Being an oil, the mixing and evaporation of fuel components into the gas phase is significantly different from a carbureted system, which uses gasoline or other low molecular weight fuels. The significance of the liquid fuel spray cannot be overestimated, since it strongly affects the pattern of air pollutants formed in the diesel system.

Another difference is the air-to-fuel ratio in the diesel combustion chamber. This air-to-fuel ratio varies spatially throughout the combustion zone due to the impact of the fuel spray. Air swirl will also influence the geometry of the flame pattern. Liquid fuel spray, air-to-fuel ratio, and air swirl interact under the high pressure regimes to influence the combustion contaminants associated with diesel emissions. As the jet of fuel is injected into the combustion chamber, the high temperature and air swirl cause the formation of a fan-shaped pattern of evaporating fuel droplets and vapors.

Consequently, four major regions have been identified in the combustion step in the compression-ignited system where air contaminant generation varies significantly. These regions are the fuel spray edge, the flame zone, the core, and the droplet or impingement zone. These four zones are illustrated in Figure 7.8.

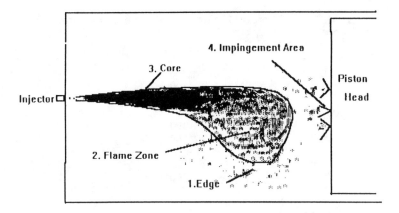

Figure 7.8. Diesel fuel spray pattern.

At the edge of the spray, the air-to-fuel ratio is too lean for flame propagation and good combustion due to the high excess air. This is a zone of formation for carbon monoxide and other products of incomplete combustion (PICs), as well as gaseous hydrocarbon fuel fragments. At low or idle conditions, this zone is relatively large and therefore emissions of hydrocarbons are greater. However, as the pressure and temperature increase with increasing load, this zone decreases, and the overall emission of hydrocarbons, CO, and PICs will decrease.

The flame zone is operating at near stoichiometric conditions at the highest flame temperatures. Therefore, this area tends to form high quantities of oxides of nitrogen but very little CO or hydrocarbons. Also, this zone is relatively long-lasting and therefore more NO_x is generated.

The spray core is that zone in which droplet evaporation is the predominant mechanism. The combustion in this zone is limited, due to the relatively slow diffusion of combustible vapors from droplets into the available surrounding air mass. Due to the diffusion-controlled nature of this combustion, the amount of swirl air interacting in this zone will significantly impact the pollutant mix.

At low loads, where a relatively smaller amount of fuel is available, some oxides of nitrogen will be formed here due to the amount of excess air available to the diffusion flame. At higher loads, however, this zone will be responsible for CO, hydrocarbons, and PICs, as well as *soot or carbonaceous particles*. Soot particulate is a significant problem with diesel fuel.

The fourth zone is that location where large fuel droplets cause most of the pollutant generation. Relatively larger droplets occur at the end of the

Table 7.3.   Typical IC Engine Pollutant Concentrations

Contaminant	Cruise Mode	Idle Mode
NO$_x$, ppm		
Diesel	1,400	50
Otto	500	100
HC, ppm		
Diesel	100	200
Otto	6,000	4,000
CO, ppm		
Diesel	500	150
Otto	60,000	10,000
Soot/PM10		
Diesel	High	High
Otto	Low	Low

injection close to the injection port. These large drops form due to reduced pressure at the end of the injection and the higher combustion chamber pressure.

These relatively large drops are also responsible for diffusion controlled combustion and likewise form soot particles upon evaporation of volatile components. Hydrocarbon and PICs are also in the emission. Secondly, some of these large drops can impinge on the cylinder head, yielding soot formation and carbonaceous deposits.

Improvements in diesel particulate emission center around increasing the chamber and nozzle design to give better swirl control and better spray patterns. Fuel injection improvements include higher pressure pumping, better timing of the injection and "rate shaping" in which an initial small charge is injected to begin the combustion followed by the bulk of the fuel at high pressure.

Pollutant Patterns

A comparison of the pollutant patterns of gasoline (Otto) engines and diesel engines is shown in Table 7.3 for two operating modes. In this table, we see the differences in emission patterns under idle conditions and under normal cruise conditions for the two engine types.

For oxides of nitrogen under cruise conditions, a diesel cycle system

Table 7.4. Influence of Trip Cycles on HC Emissions[a]

Condition	Trip Length		
	5 Miles	10 Miles	20 Miles
Cold Start	9	9	9
Cool Down	2	2	2
Running Exhaust	3	6	12
Total HC, grams	14	17	23

Source: California Air Resources Board, 1992
[a]Grams total hydrocarbons.

produces significantly higher overall NO_x emissions than a spark-ignited or Otto cycle engine. At idle conditions, as noted above, the NO_x emissions are much lower than for the spark-ignited systems.

With respect to hydrocarbons and carbon monoxide, the spark-ignited emissions are significantly higher, in some cases by an order of magnitude, than diesels, due to the excess air and compression conditions noted earlier for diesels. Carbonaceous particulate formation for diesels is significantly greater than for spark-ignited systems, due to the higher molecular weight and oily nature of diesel fuels.

Hydrocarbon Emissions from Trip Cycles

Quite apart from the comparisons of the two engine types is the influence of cold start and hot soak hydrocarbon emissions as opposed to those from running exhaust. In Table 7.4, the influence of trip length on hydrocarbon emissions is summarized. *Cold start* emissions are fairly constant at about nine grams of fuel hydrocarbons per start. The end-of-trip period or "hot soak" generates roughly two grams of hydrocarbons for the average car equipped with a catalyst. Cold start and hot soak emissions are termed the *standing emissions*. The running exhaust emissions are approximately 0.55 grams per mile and are thus a function of the total miles traveled.

The significance of these differences in running versus standing emissions is in the control approach taken. Implementing emission reductions in the first two to three minutes of operation, as well as during the cool-down at the end of the trip would significantly lower overall air quality impacts. Thus, air quality management strategies will have to identify techniques for controlling evaporative hydrocarbon emissions as a function of the *number of individual trips*.

These standing versus running emissions appear to explain some studies which show greater frequencies of elevated ozone levels on weekends when, presumably, there are a greater number of short trips but a lesser number of total miles traveled in areas such as Los Angeles.

IC ENGINE EMISSION CONTROL OPTIONS

The two major approaches to minimizing emissions from internal combustion engines are either through changing the operating conditions or changing the design of the engine itself. These function apart from changes in fuels composition, which are handled separately.

The former approaches have received significant attention in order to minimize emissions prior to back-end, or tailpipe, control technologies.

Effects of Operating Conditions

The single most important impact on combustion emissions is due to the air-to-fuel ratio. Figure 4.5, seen earlier, is an illustration of pollutant concentrations versus air-to-fuel ratio for gasoline combustion. The *equivalence ratio* is the relationship between the fuel-to-air ratio of the operating system to the fuel-to-air ratio at stoichiometric, or ideal, conditions. An equivalence ratio less than 1.0 indicates lean conditions (excess air). An equivalence ratio greater than 1.0 indicates fuel-rich conditions.

In fuel-rich regions, hydrocarbon and CO emissions tend to predominate and reach their *minimum* on the slightly lean side of the stoichiometric ratio. Oxides of nitrogen tend to *peak* on the lean (excess air) side of stoichiometric mixtures. It should be noted that at very high air-to-fuel ratios, CO and hydrocarbons again increase, due to the temperature quenching effect of excess air.

There is a "balancing act" performed between combustion controls for NO_x and those for hydrocarbons/CO. The maximum combustion temperature curve is similar to the NO_x curve. Thermal NO_x emissions are highly temperature-dependent.

Spark Timing

For spark-ignited ICEs, the influences of changing the spark timing may be significant. If the spark timing is advanced, overall temperatures during the cycle tend to go up, and NO_x concentrations will therefore increase. Advancing the spark means that the spark occurs earlier than normal during the compression cycle. Retarding the spark tends to lower oxides of nitrogen but may only marginally reduce hydrocarbon emissions. The effect of spark timing and equivalence ratio is seen in Figure 7.9 on oxides of nitrogen concentrations.

Compression Ratio

The effect of compression ratio will significantly increase both oxides of nitrogen and hydrocarbons in direct proportion. Figure 7.10 shows the hydrocarbon concentrations as a function of compression over a range of equivalence ratios. In Figure 7.11, the influence of compression on oxides of nitrogen concentrations is apparent. The effect is directly proportional to increasing compression ratios.

The major effect on NO_x is in the excess air or lean regions. For hydrocarbons, the effect is "across the board." The hydrocarbon trend is reflective of the variation in thickness of the quench zone or boundary layer in which fuel gases are contained.

At higher compression ratios, a greater quantity of fuel would exist in that boundary layer and would therefore be exhausted with the exhaust stroke. The effect on oxides of nitrogen is clearly one of increased temperatures in the chamber with increasing pressurization.

Engine Speed

The effect of engine speed, or rpm, on emissions is variable. An increase in engine speed tends to reduce hydrocarbon concentrations due to an increase in turbulence in the combustion chamber. Thus, a greater percentage of the hydrocarbon fuel fractions near surfaces become entrained in the bulk gases and are burned. Likewise, exhaust gas temperatures are increased with speed, and that promotes further combustion and reductions of CO and hydrocarbons (if the mixture is lean). This effect, however, causes chamber temperatures to increase and oxides of nitrogen concentrations therefore also increase with engine speed.

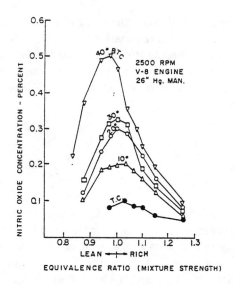

Figure 7.9.   Effect of spark timing and mixture ratio.

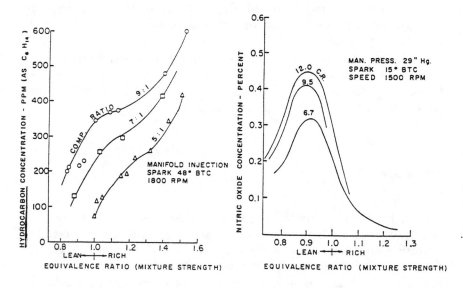

Figure 7.10.   Influence of compression ratio and mixture ratio on hydrocarbons.

Figure 7.11.   Influence of compression ratio and mixture ratio on nitric oxide.

Engine Power

The effect of engine power output is a direct function of the amount of fuel being injected into the cylinder. Therefore, the effect of power has an impact on the fuel-to-air ratio. As power is increased, the fuel mass increases through the engine, and the air-to-fuel ratio decreases. This tends to increase both hydrocarbons and CO. Due to higher temperatures from the increased amounts of fuel, NO_x concentrations increase with load until a rich condition occurs (as under high acceleration demands). At that point, NO_x begins to decrease.

Engine Temperatures

Overall engine temperatures, as measured by the temperature of the cylinder surface walls, also have an influence. Operating in hot ambient air or operating in conditions which cause the engine coolant temperature to increase would increase oxides of nitrogen and decrease hydrocarbons. In one experiment, an increase of the engine surface temperature by 75° generated a 73% increase in overall NO_x emissions. In another experiment, an increase in coolant temperature of 100°F decreased hydrocarbon emissions by one-third.

Engine Cleanliness

As mentioned earlier, the impact of deposits will have a significant impact on carbon gas emissions. Therefore, a cleaner or newer engine will have a positive impact by lowering carbon gas emissions. Deposits serve as "sponges" for fuel fragments (hydrocarbons) which are emitted later in the cycle. On the other hand, cleaner engines are responsible for higher temperatures, and higher NO_x emissions result.

Design Influences on IC Engines

Engine design changes modify the combustion process, which has an effect on air contaminant emissions. By lowering the surface-to-volume ratio of the combustion chamber, more of the hydrocarbons will be in the bulk gas and therefore, the combustion will be more complete. Thus, designing the combustion chamber to reduce this ratio to as low as possible (considering cost, design, feasibility, and reliability) will improve combustion, thereby lowering emissions of hydrocarbons.

The influences of exhaust back pressure and valve overlap affect hydrocarbons and NO_x emissions by influencing the amount of exhaust gases remaining in the cylinder following combustion. Increasing engine back pressure and overlapping the valve timing will cause a greater quantity of the last stroke's gases to be retained in the cylinder. This causes the oxides of nitrogen to decrease due to lower temperatures (a function of the greater quantity of residual non-combustibles), and lower air-to-fuel ratio. Hydrocarbon emissions tend to increase, however, as the air-to-fuel ratio is reduced.

A similar approach, but one which is much less complex than using back pressure and valve timing, is the inclusion of an *exhaust gas recirculation* system. The recirculation of exhaust gas significantly lowers oxides of nitrogen emissions but causes only minimal changes in CO and hydrocarbons. This is due to the lower gas temperature (from higher inert gas concentrations) and the lower oxygen content of the mixture.

Exhaust gas is recirculated back into the combustion air intake manifold by virtue of a control valve. It has been found that a 5% gas recirculation may reduce uncontrolled NO_x emissions by more than 50%, depending on other engine parameters. Up to 15% recirculation may cause controlled NO_x emission reductions of more than 75%. Above this level, CO and hydrocarbon emission tend to increase. This approach is nominal in cost and complexity. At full load conditions, or maximum acceleration, exhaust gas recirculation is bypassed due to the increased power requirements and the lower air-to-fuel ratios caused by heavy acceleration.

Fuel injection lowers hydrocarbon emissions by allowing more precise fuel control, particularly under cold start conditions and deceleration. This reduces the need for fuel enrichment during start-up and load change, as compared to carbureted fuel systems.

Computerized engine controls, coupled with diagnostic features, provide for optimized operation of the combustion process due to fuel-to-air control, spark timing, and enhanced control settings for engine control devices, such as exhaust gas recirculation valving. This optimization is expected to minimize hydrocarbon and CO emissions as well as oxides of nitrogen.

Two-Stage Combustion

Without totally redesigning the internal combustion engine, it is possible to take advantage of certain aspects of the air-to-fuel emission curves so that low NO_x emissions (characteristic of a fuel-rich mixture) can be coupled with the low hydrocarbon and CO emissions in the excess air regions of the fuel-to-air ratio. This has led to the redesign of the top of the cylinder so that the combustion chamber consists of two parts. Figure 7.12 illustrates this two-stage combustion engine, also called a *stratified charge*.

In this system, the combustion takes place in two phases. There is a fuel-rich chamber on the top and a fuel lean portion in the cylinder itself.

In the fuel-rich combustion chamber, approximately 40% to 70% of the combustion air is supplied along with the full charge of fuel. In this fuel-rich combustion, very low oxides of nitrogen are formed; however, large quantities of carbon monoxide, PICs and fuel fragments occur. The rich hot combustion gases then pass through a restricted opening into the lean portion of the cylinder.

In the lean portion of the process, the balance of the required air is supplied, which causes the overall combustion process to operate in the lean fuel-to-air regions such that burnout is completed. The exhaust gases then drive the piston down, providing power to the system. CO and hydrocarbons are consumed, and the oxides of nitrogen are minimized due to the operations at higher excess air, which promotes a lower overall combustion temperature.

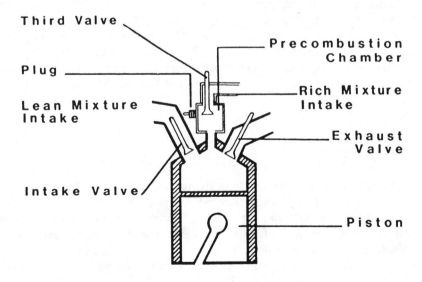

Figure 7.12. Stratified charge engine.

EXTERNAL CONTROL APPROACHES

The original approach to automotive emissions control has been to literally hang a control device on the tailpipe or other sources of fugitive emissions. This approach has given way to a combination of options, including engine redesign, engine operations and fuel composition changes, along with control devices. As more research is performed to accurately gauge emissions from tailpipes and fugitive sources, pollution control devices will both remain and be enhanced in their effectiveness.

Fuel Recapture Systems

Evaporative emission controls are fairly simple and inexpensive, since they consist of an activated carbon adsorption unit with ducting from the sources of evaporative emissions (fuel tank) to the adsorber.

Emissions of hydrocarbons during displacement by refueling have been addressed by Stage II vapor recovery systems at gasoline-dispensing facilities. These include the rubber boot and the vapor-return line which draws gasoline vapors back into the storage tank. Spillage is minimized due to an interlock system which senses a slug of liquid entering the vapor line and stops the flow of fuel into the tank.

Activated charcoal canisters to collect fuel transfer emissions, in addition to vapor recovery are mandated by the Clean Air Act Amendments. A "bottom fill pipe" on older models served to eliminate the splashing and frothing action during filling which increases hydrocarbon emissions.

Blowby gases coming from the engine are typically recirculated back into the combustion air intake and serve to return hydrocarbons to the combustion chamber.

Catalyst Systems

Thermal oxidizers were the first attempt to deal with hydrocarbon and CO emissions. These devices were large insulated chambers with baffles which in order to operate were raised to temperatures between 800°C and 900°C. These units were effective, provided that an additional air flow of 20% to 30% of the stoichiometric air flow was provided by an air pump. Exhaust gases that are operated at overall lean conditions showed a much lower efficiency. The key benefit of thermal systems is that they are not "poisoned" by lead, sulfur, or phosphorus contaminants in the exhaust.

Catalyst-based systems are currently tasked with the requirement of significantly reducing hydrocarbon, CO, and NO_x emissions under varying modes of operation. These come in two types: the oxidizing system and the dual stage system of catalysts.

The first attempt was the *oxidative single stage* system which dealt with hydrocarbons and CO. These single stage systems were typically high surface area cartridges containing a noble metal catalyst. Being catalytic, they could operate at lower temperatures than incinerators and still produce significant reductions in both CO and hydrocarbons.

The function of the catalyst, by using materials such as platinum and palladium, is to lower the activation energy required to oxidize the fuel and CO to CO_2. By lowering the activation energy, a lower temperature is required; however, the requirement for oxygen still remains.

An exhaust gas high in CO or hydrocarbons, (such as from a cold start or high acceleration) has poor conversion efficiency, since oxygen or air-injection systems are required. In stoichiometric or lean conditions, the catalyst may work without an air pump for CO and hydrocarbon control. Potential problems with oxidizing catalysts are that they may be sensitive to poisoning by other trace elements in the fuel or exhaust gases and, secondly, that all fuel elements will be fully oxidized. These may form materials such as sulfur dioxide and sulfuric acid aerosol. Oxides of nitrogen are not addressed in such oxidizing or single-bed catalytic converters.

Two-stage or dual-bed catalytic systems attempt to get around the problem of oxides of nitrogen emissions by providing a second stage. In these systems, the oxides of nitrogen are eliminated first by operation of a catalyst under fuel-rich conditions so that reducing gases, such as hydrogen, CO, and fuel fragments combine with NO to produce molecular nitrogen (N_2) plus some additional CO_2. As the exhaust gases pass over to the oxidation bed, additional air is injected to complete the oxidation of CO and hydrocarbon fragments. As a result, the dual-bed systems are larger, heavier, and more expensive than the single-bed systems. Other concerns are for lowered fuel economy, since the engines must be run in a fuel-rich condition.

Problems with a dual-bed system include potential poisoning of the catalysts due to the presence of other elements. Also, ammonia may be produced in the first stage which will then be more easily oxidized to NO_x in the second stage, should the reducing conditions of the first stage be too efficient. Another problem is that catalytic systems only heat up after several minutes of operation due to a cold engine. Therefore, the bulk of those emissions tends to be emitted initially without being affected, because the catalyst is not up to temperature.

Table 7.5 lists the typical emissions from a three-way catalyst-equipped

Table 7.5.   Typical Gasoline Exhaust Emissions

Pollutant	Emissions (gm/km)[a]
Total hydrocarbons	0.14
Oxides of nitrogen, total	0.38
NO	0.35
N_2O	0.031
NO_2	0.005
Carbon monoxide	1.8
Toxics, total	0.0094
Benzene	0.0069
1,3 butadiene	0.00052
Formaldehyde	0.0011
Acetaldehyde	0.00077

[a]Three-way catalyst vehicles.

vehicle. The three ways, of course, refer to the CO, NO_x, and hydrocarbon emissions addressed by these systems. They are still packaged in two-stage operating units.

Diesel Particulate Controls

For diesel fueled systems, major concerns have been for the carbonaceous soot or particulate emissions commonly associated with such compression ignition systems. Therefore, attempts to provide better emissions control for diesel systems have focused on the diesel particulate trap.

Soot particulates may be emitted at rates of 0.1% to 0.5% of the total fuel input mass by weight. Due to the low density and small size of carbonaceous soot, regular filters are not appropriate, since they plug quickly. Two approaches have been tried for the diesel particulate controls. These include oxidizing ceramic packing filters, which attempt to burn off the carbonaceous soot during operations, or change-out filters which must be periodically regenerated. Research is continuing on such diesel particulate filter systems.

There are some indications that eventually all catalyst systems may be heated prior to the start of the engine, since engines under cold start conditions emit a high level of air contamination. Other possibilities are to mount the catalyst very close to the engine so that radiative cooling will be minimized.

FUEL CHANGE EFFECTS

Since combustion in internal engines is a kinetically-driven system, the composition of the fuels themselves may significantly impact the emissions. The major benefit of fuel changes is that all vehicles on the road would be using lower emission gasolines; therefore, emissions improvements would begin immediately rather than over a 5 to 10 year period for newer, lower emitting engines to replace the older ones.

In general, fuel changes have been required to lower the volatility and evaporation rate of the fuel, and to provide cleaner-burning gasolines which will lower engine deposits. A more recent emphasis has been to lower air toxic emissions (such as benzene), and to lower the percentages of those components of gasoline which are photochemically reactive (those which promote greater ozone production per unit mass). Figure 7.13 indicates the photochemical oxidation potentials for different classes of organic compounds which contribute to such photochemical ozone production.

A recent study, termed the "Auto-Oil" program, has indicated the potential emission reductions that can be attained by changing fuel compositions.

In the Auto-Oil program, the effects of changing the compositions of five different parameters (olefins, aromatics, 90% distillation temperature, percent sulfur, and oxygenates) were measured on the exhaust emission species. The exhaust emission species were nonmethane hydrocarbons, carbon monoxide, NO_x, benzene, 1-3, butadiene (1,3 Bd), formaldehyde (HCHO), and acetaldehyde (C_2H_4O).

The effects of *reducing olefin content* from an average of approximately 20% in the test gasoline to 5% resulted in the emissions changes noted in Figure 7.14. In this study, no significant changes were found in CO, formaldehyde, acetaldehyde, or benzene, but the 1-3, butadiene fractions were reduced by approximately one-third. In addition, oxides of nitrogen had a slight reduction of about 6%. The overall total mass of nonmethane hydrocarbons (primarily fuel fragments) increased by approximately 7%. Thus, reducing overall olefin content of gasoline could provide some decreases in ozone potential by virtue of the 1-3, butadiene emission decreases.

Reductions of the aromatic content from 45% to 20% led to dramatic *decreases* in benzene (45% reduction), CO, and nonmethane hydrocarbons (at about 12% each), with no significant increases in NO_x. While effective in reducing the above, however, this fuel change resulted in emission increases of formaldehyde, acetaldehyde, and the ozone-forming 1-3, butadiene.

Reducing the "high boiling fractions" (T_{90} = 90% distillation temperature) in gasoline, as seen in Figure 7.15, produced significant reductions in all carbon species except for CO, which showed no significant change.

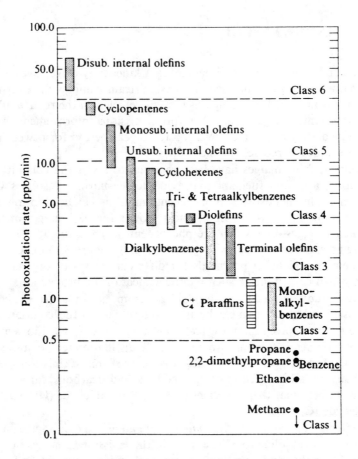

Figure 7.13. Hydrocarbon reactivity as indicated by the rate at which NO is oxidized.

Oxides of nitrogen indicated a nominal increase of about 5%. This change was accomplished by reducing the boiling point of the 90th percentile fractions of gasoline from 360°F to 280°F.

Probably the most significant emission reductions due to composition changes average fuel sulfur content from 450 ppm to 50 ppm, as seen in Figure 7.15. In this experiment, all contaminant emissions were reduced significantly except for 1–3, butadiene, which showed no effect and formaldehyde, which showed a 45% increase.

The purpose of adding oxygenated compounds (MTBE or methyl t-butyl ether) to gasoline is to force lower CO emissions. To some degree the

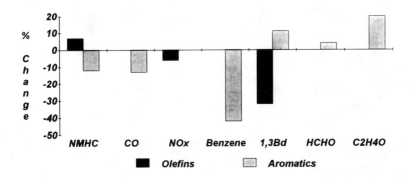

Figure 7.14.   Fuel change effects, olefin and aromatic HC reductions.

addition of oxygenated fuel components to gasoline also helps to provide better "anti-knock" characteristics to engine performance. Figure 7.16 from the Auto-Oil Program experimental studies indicates the change in emissions by using a 15% oxygenated gasoline. No significant changes were found in oxides of nitrogen, benzene, or acetaldehyde. Reductions of CO and nonmethane hydrocarbons were found, as well as reductions in 1,3-butadiene. Formaldehyde emissions, however, increased by approximately 27%.

One further indication of the Auto-Oil data is that as the volatility of the gasoline was reduced (as measured by the Reid vapor pressure), a one pound per square inch in RVP caused a reduction of 46% in the evaporative hydrocarbon emissions during fuel storage in the system.

As a result of the Clean Air Act Amendments and similar studies, California has mandated the gasoline composition requirements seen in Table

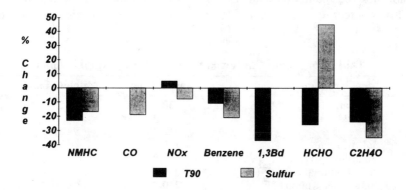

Figure 7.15.   Fuel change effects, T90 and sulfur reductions.

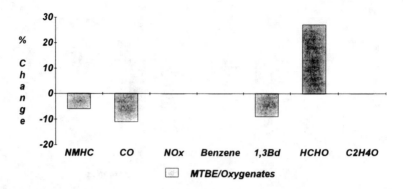

Figure 7.16.   Fuel change effects, 15% oxygenated compound addition.

7.6 for March 1996. Significant among these are changes in the overall average composition of the fuel mandated by state regulations. Should other states opt into the "California approach," these fuel requirements could become more widespread.

Diesel Fuels

Composition changes for diesel fuel focus on sulfur content, aromatic hydrocarbon content, and particulate emissions. The latter are addressed by oxidative particulate traps and by lowering the ash or mineral content of diesel fuels.

In addition to mandating lower sulfur contents, regulations increasingly are focusing on lowering the aromatic content of diesel fuels. Figure 7.17 indicates the mandated changes for California which will lower the fuel sulfur content to a maximum of 500 ppm (0.05%) and a maximum aromatic

Table 7.6.   1996 California Gasoline Composition Limits

Component	Allowed Now	Absolute Limit
Aromatics[a]	32	25
Olefins[a]	9.9	6
Benzene[a]	2	1
Sulfur, ppm	150	40
Volatility, psi	7.8	7
90% Boiling point, °F (T90)	350	300
50% Boiling point, °F	220	210

[a]% by volume.

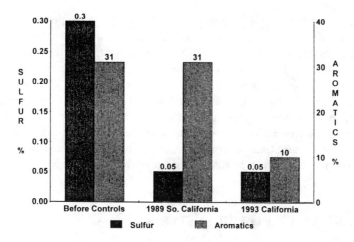

Figure 7.17.   California regulations require cleaner fuel.

content of 10%. These mandated cleaner diesel regulations force an approximately 84% reduction in total sulfur emissions and approximately a 68% reduction in aromatic hydrocarbon emissions.

Alternative Fuels

Other options to reformulated gasoline and diesel are the oxygenated fuels, methanol, and ethanol. The lighter end components of crude oil, such as liquid petroleum gas and compressed or liquefied natural gas, are also viable contenders for the fuel of the future.

Pure methanol fuels were expected to have significantly lower evaporative emissions and running losses. However, one problem associated with methanol is the difficulty in starting the engine. Overall, photochemical modeling studies indicate that methanol has a lower ozone-forming potential than gasoline; however, methanol has high emissions of formaldehyde. Formaldehyde emissions from the current generation of methanol vehicles is on the order of 5 to 50 mg per mile, or about an order of magnitude higher than those from gasoline. Other problems are that both methanol and formaldehyde are *hazardous air pollutants* and represent increased risks to public health.

Ethanol has concerns similar to methanol due to its vapor pressure at low temperatures and the formation of acetaldehyde during combustion. Ethanol and acetaldehyde both present concerns for adverse impacts on public health due to their toxicological effects. Ethanol is an expensive fuel, since it is made from fermentation processes and then distilled into a fuel.

For liquefied petroleum gas (LPG), the nonmethane tailpipe emissions can be relatively high; however, there appears to be a significant reduction in ozone-forming potential for LPG relative to gasoline. Olefin emissions, however, appear to be somewhat higher than for current gasolines. Compressed natural gas appears to be the best candidate from an emissions standpoint since the combustion kinetics can be more easily optimized (because it is essentially a one-component fuel); however, there are some concerns for safety and availability of sufficient fuel outlets.

ALTERNATIVE POWER SYSTEMS

Other means of providing power to the automobile include hydrogen fuel, solar power, and fuel cells.

Hydrogen is currently made totally from natural gas and steam via a water-gas reaction. Using hydrogen in cars would generate few reactive hydrocarbon emissions. However, since these experimental power plants burn air, the oxides of nitrogen emissions may be significant. Likewise, the production of hydrogen has emissions associated with it at the production facility; therefore, hydrogen fuels may merely shift direct emissions from combustion to the hydrogen production plant.

Fuel cells have the potential to fill a niche since they produce DC current directly from a low temperature "combustor." This current drives an electric motor. However, the power density of fuel cells is limited.

A purely electric auto propulsion system appears to be an attractive long-term transportation power system; however, it suffers from two major deficiencies. First, the electrical power must be generated at a power plant, which has its own associated air pollutant emissions. Secondly, the power must be stored in a battery system. Most of today's research centers around attempts to provide energy storage in a battery system to provide power and range comparable to that seen in internal combustion engines.

Figure 7.18 indicates the energy and power ranges of potential power systems for automobiles. In the near term, it appears that alternative fuels may hold the best hope for lowering emissions. In the long run, the limitations on electrical systems center around the search for batteries, sufficient energy storage and power density to come close to IC engines. Until both power and energy densities for alternative systems are found to approximate that of IC engines, ICEs will remain the propulsion system of choice, using some form of fossil fuel.

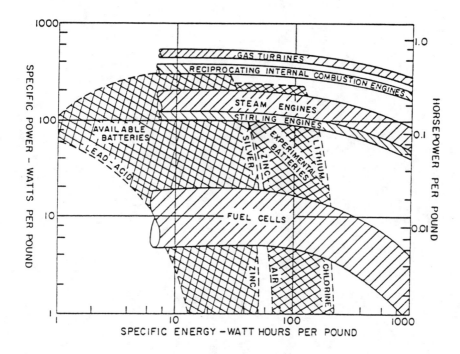

Figure 7.18.   Comparative performance of various power systems.

8   GLOBAL CONCERNS

When you first look into climate change, you realize how little you know. The more you look into it, the more you realize how little anyone knows.

Dr. Ralph Cicerone,
University of California, Irvine, March 1992

The earth and its atmosphere are a dynamic system. Meteorology, emission sources, air pollution control strategies, ocean temperatures, volcanoes, and sunspots, as well as their second order effects, make for a system too complex to truly understand. In probably no other area of air quality management is there greater uncertainty than in global issues. With regard to anthropogenic air pollutant impacts on acid rain, stratospheric ozone changes, and climate change, facts are few but opinions are many.

Popular opinion concerns for acid rain deal with the potential vegetation effects; stratospheric ozone concerns are for potential surface UV light penetration and resultant fears for cancer; climate change concerns are for potential rising sea level and impacts on agriculture. In each case we know some things well, primarily *data measurements* of relevant parameters, such as temperatures and gas concentrations at various points and times. In other cases, the best we have are theories and computer models. The concerns are many, and the potential costs are high. This is not unexpected since the impacts of any air quality management approach will potentially affect virtually every area of society. Implications and fears are many: agriculture, health, economics, business, and international relations are just some of the areas which have a stake in air quality management for global issues.

It is certain that no one has all the answers. Indeed, it will take decades to determine actual trends in the atmosphere.

The Challenge

It is appropriate to speak of *change* in measurable parameters, since these are measurable scientifically. However, terms such as *loss* or *fluctuation* reveal different levels of knowledge, or more appropriately, presuppositions. When one speaks of change, we can say that there are mathematical differences in measurable parameters versus time. Thus, when we measure differences in measurable parameters, we may base our conclusions on the scientific method, because these are based upon evidence.

Terms such as *loss* imply an irrevocable or irretrievable diminishment in some quantity which may not be verifiable at the present. *Fluctuation* denotes a dynamic process over time which may be a better term to use when dealing with changing data whose true *cause* is unknown at present.

The challenge is to evaluate changes accurately without becoming advocates. One goal is to be fully cognizant of the accuracy of our measurements. Those parameters which may only be modeled, estimated, or assumed are based upon presuppositions. An open-minded appraisal of measurable facts is the best approach. At all times researchers need to be fully cognizant of the uncertainties and potential discrepancies in such models as new evidence becomes available. The reason that one must be careful of the information which models yield is that they are, at best, *approximations*-based limited data sets. Therefore, small changes in input data, factors, or other "constants" may produce significant changes in modeled output.

Data and Records

With respect to scientific measurements, there is only a very limited period of time during which accurate real time measurements of the physical world have been taken. For some parameters the maximum time period over which we have accurate measurements is about 150 years (pH). In other cases, such as the *existence of methane,* our knowledge dates back barely 200 years. Therefore to evaluate potential air quality management strategies for their global impact, one must take into account other evidence available for time periods prior to "real time" scientific measurements. This allows one to put modeled calculations into perspective and maintain a sense of proportion.

Historical records are the first source of information. As an example, historical accounts of climate experienced by various population groups and their influences on history allow us to make general statements relating to climate, such as the "little ice age" which peaked in Europe in the late 17th century. Likewise, agricultural patterns may allow us to see the general trend of climate in a location such as North Africa. During the height of the Roman Empire, North Africa was considered the "granary" of the Empire, since wheat was grown throughout that region. This indicates that there once was a much wetter climate in that area than at present. Recent NASA satellite observations show river channels with complete riverine tributary systems buried beneath the sands of the Sahara desert, thus verifying greater rainfall in the past.

Indirect evidence, such as tree rings or gas compositions of micro bubbles in ice cores may or may not point to climate changes as well. It is important to realize that *interpretation* is somewhat open for indirect records. The degree of accuracy of indirect evidence compared to present levels of instrumentation and real time data is unknown; however, these records do allow for qualitative trend analyses over periods of centuries.

Our focus must therefore be on what is known to be true first, based upon observable facts, second on historical records, and finally on indirect evidence. From this information, one may develop models, "what if" scenarios and alternate views of the same data. From this different societal management, options may be developed. In all events, we must be aware of the uncertainties in any approach beyond that which is verifiable by measurement techniques.

ACID DEPOSITION

All rainfall is acidic. Pure water, which has a neutral pH of 7.0, will, because it is the universal solvent, dissolve some of the gases next to its surface. Thus, raindrops forming from condensation nuclei will have some dissolved nitrogen, dissolved oxygen, and all of the gases noted in Chapter 1, including carbon dioxide.

With the exception of oxygen, nitrogen, ammonia, hydrogen, and the noble gases, atmospheric gases have an acidic property. That is, when they are dissolved in pure water, they will lower the pH into the acid region (pH less than 7). Pollutant gases, such as sulfur dioxide and chlorine will also form acids when dissolved in pure water.

Water Plus Air

One of the major considerations when discussing the effect of gases dissolved in water is the chemistry of the droplet itself. When there is a dissolution of a gas such as carbon dioxide in water, we find not only the dissolved gas, carbonic acid (H_2CO_3), but also an equilibrium between the dissolved gas and its ionized form. For CO_2, a bicarbonate ion and a free proton are generated. Protons give water its acidic characteristics. The pH in this case is 5.6, or well into the acid range. Equation 8.1 (dissolution), therefore, gives rise to Equation 8.2 (dissociation):

$$H_2O + CO_2 \rightarrow H_2CO_3 \tag{8.1}$$

$$H_2CO_3 \rightarrow H^+ + HCO_3^- \quad (pH\ 5.6) \tag{8.2}$$

Where other gases (such as ammonia) are present, other reactions are possible. If the acid protons are neutralized by an alkali (Eq. 8.3), the bicarbonate will react further to yield a second proton and a carbonate ion:

$$HCO^-_3 + NH_4OH \rightarrow NH_4^+ + CO^=_3 + H_2O \tag{8.3}$$

On the other hand, if more protons are added, the equation shifts back to give more carbonic acid which will be free to liberate gaseous CO_2. The overall effect is a *buffering action* which acts to keep the pH from changing drastically.

Water Plus Soils

Probably the most significant element of the debate is the influence of acidic water once it reaches ground level. This does not take into account acid fogs, which are a different end product but which may have health effects of their own. Once rainwater reaches soil, the key element in that new matrix is the relative percentage of minerals in that soil. In particular, the relative abundance of calcium, magnesium, and, to some extent, aluminum ions in the soil structures largely determine the pH of the water therein.

The alkaline earth minerals (calcium, magnesium) act as basic compounds to present another buffering action to any acid elements in deposited rain. Thus, an effect such as Equation 8.4 may be seen:

$$CaCO_3 + 2H^+ \rightarrow Ca^{++} + H_2CO_3 \tag{8.4}$$

In this case, the hydrogen ions are neutralized by calcium carbonate (a typical component of many soils and rock formations) to yield the calcium ion plus carbonic acid dissolved in the water. The relative abundance of a number of minerals in the soil structures has a large effect on pH. The

Table 8.1.   Median Variable Ion Concentration (mg L⁻¹) 1979–84 Atmospheric Deposition Program (NADP) Sites

Ion	Lamberton Minnesota	N. Atlantic Lab, Massachusetts	Kane Pennsylvania
pH	6.00	4.67	4.27
SO_4^{2-}	1.88	1.54	3.48
NO_3^-	1.74	0.73	2.08
NH_4^+	0.81	0.08	0.28
Ca^{2+}	0.49	0.13	0.16

Source: A.H. Legge and S.V. Krupa, *Acidic Deposition: Sulphur and Nitrogen Oxides,* Lewis Publishers, Chelsea, MI, 1990.

effects of mineral content, as well as *humus and organic acids from plant decay*, will reduce soil water pH to between 4.5 and 5.5.

One would expect, therefore, that the more granitic the soil and the fewer dissolved alkali minerals present, the more likely it is that precipitation pH values will be unbuffered and, therefore, more likely to show acid water levels in the pH 4.5 to 5.5 range. Greater concentrations of alkaline ions significantly buffer or neutralize acid components originating in rain.

Acid Rain Studies

Concerns exist for receptor areas downwind of major anthropogenic sources of acidic gases, such as sulfur dioxide. A number of studies have been performed in the last few decades reviewing the entire field of anthropogenic acid gas emissions and receptors such as lakes and streams in North America.

Analyses of early studies (1964 through 1977) on acid deposition indicated, among other things, that:

1. The annual pH of precipitation showed no long-term significant change from 1964 to 1977;
2. A linear regression of data points indicated no statistically significant trends in H^+ deposition; and
3. There has been a decrease in $SO_4^=$ since 1964, but an increase in NO_3^- over the same time period.

Over 80 sites were studied for the 1979–1984 period for pH, H^+ strength, ion concentrations, and precipitation in the northern United States. Table 8.1 summarizes the median concentrations at three of those sites representing a range of receptor locations with differing upwind air pollution source strengths. From this study it appeared that the pH and calcium concentrations are directly related — when the pH is high, the calcium ion strength is high, and vice versa. No obvious correlation appeared between pH and either sulfate, nitrate, or ammonium ions.

A different study (1986) found the following conclusions: (1) The eastern half of the United States experiences ion concentrations of SO_4 and NO_3 that are greater by a factor of 5 than those levels found in remote parts of the world, and (2) data on the chemistry of precipitation before 1955 should not be used for trend analysis, primarily due to the difficulties in establishing a correlation with *present methods* of measurement and those used previously.

Other investigators indicated that a pH of 5.6 may not be a reasonable reference value for unpolluted precipitation pH. Some have questioned the validity of using pH 5.6 as the background reference, due to naturally occurring acids as responsible for lower pH values of rains in certain areas. Likewise, the *times* during which the rain was collected varied in pH due to the scavenging efficiencies of rainfall and the times between storms. pH values of rainfall between 4.5 and 5.6 may be due to those variabilities alone.

The NAPAP Findings

The most recent study, the ten-year, $500 million, National Acid Precipitation Assessment Program (NAPAP) was completed in 1990 and then extended under the Clean Air Act Amendments. NAPAP found some significant but similar trends and effects. NAPAP employed 700 of the world's top aquatic, soil, air, and atmospheric scientists in an exhaustive study as to the effect of acid rain on receptor areas as a result of anthropogenic emissions.

As a part of the NAPAP program, EPA scientists performed an exhaustive study of a correlation between acid precipitation and acidity levels in various receptor waters. Table 8.2 summarizes the results of that study for two areas which were suspected of experiencing the highest impact due to being in receptor locations of high sulfur dioxide emissions. These two sites were the northeastern United States and the southern Blue Ridge province in the Appalachian mountain region.

Of the five factors which were investigated for their correlations to surface water acid neutralizing capacity, acid rain had virtually no correlation in either location. The highest correlation factor was to the receptor soil's chemistry, followed by weaker correlations to depth to bedrock (alluvial

Table 8.2.   NAPAP Acid Deposition Correlation Summary: Area Factors

Factor	Northeast U.S.	S. Blue Ridge
Soil chemistry	Fairly strong	Fairly strong
Depth to bedrock	Moderate	Weak
Geology	Weak	Moderate
Land use	Weak	None
Acid rain (SO_2)	None	None

Source: EPA, National Acid Precipitation Assessment Program Summary, 1991.

materials) and geology (strata). Land usage had only a weak correlation in the northeast and no correlation in the southern Blue Ridge area. Acid rain had no correlation in either case.

Other evaluations performed during the NAPAP study indicated that Ohio, which had the highest acid gas emissions in the entire United States, had virtually no acidic lakes or streams. On the other hand, Florida, which has one of the lowest levels of acid deposition, has the highest percentage of acidic lakes in the nation at 20%. These studies indicate that while acid gases such as SO_2 have something to do with aquatic acidity, they are the least influential factor studied.

As another aspect of the NAPAP study the EPA evaluated pH measurements taken in lakes over the last 140 years. These included a comprehensive core sediment analysis of all acidic lakes (lakes with a pH less than 5.5) in the Adirondack mountains. This evaluation indicated that not only were 90% of them acidic in 1850, but the average acidity today is virtually unchanged from pre-industrial times. Table 8.3 indicates the changes of pH over time in the Adirondack mountains and in the Florida acidic lakes. These are grouped into those which have a pH of less than 5.5 and those which had a pH of less than 5.0. The apparent pH change in all cases was lower by 0.35 pH unit or less. However, it should be noted that the standard

Table 8.3.   NAPAP Historical Assessment: Lake Acidity vs. Time

Lake Categories	1850 pH	1986/88 pH	pH Shift	Actual Shift[a]
Adirondack Acid Lakes				
pH < 5.5	4.95	4.78	-.17	None
pH < 5.0	4.78	4.63	-.15	None
Florida Acid Lakes				
pH < 5.5	5.11	4.81	-.30	None
pH < 5.0	5.00	4.65	-.35	>S.E.

[a]Standard Error (S.E.) of pH measurement = 0.3 pH units.

error of the measurement pH was ±0.30 pH units. Except for the most acidic Florida lakes, there was no statistical change in pH levels measured 140 years ago from those measured today.

One of the key factors which occurred over this period of time was a *change in land use patterns and widespread forest clearing*. In the mid 19th century, forests covered many of these areas, and the surface water pH values were relatively low (i.e., approximately 5.0). Over the course of approximately 50 to 60 years, the land was cleared. In each of these cases, the soil chemistry changed due to changes in the naturally occurring acidic conditions in soil structures. Fires raise soil alkalinity by replacing acidic forest floor organic compounds with alkaline ash materials. These include the cations calcium, magnesium, and aluminum. Fires release them from the soil matrix so that they more quickly and easily neutralize naturally acidic rain (approximately pH 5.0). Other soil analyses show that clear cutting of fir forests raises the pH of the soil from 5.0 to 7.0, and slash and burn fires raise it from 4.95 to 7.6.

In 1986, the National Research Council commented in a paper that core sediment analyses suggest that acidic lakes were relatively common in the Adirondack mountains and in New England before the Industrial Revolution. Woods Lake in that region of New England has a current pH of approximately 4.9. That is more acidic than the pH 5.6 found in 1915, but practically unchanged from the 1850's value of pH 5.0.

In other words, lakes have been returning to their *natural acidic state* from the temporarily more alkaline condition during the period from 1900 to 1940 due to clearing and burning.

Is there no effect due to acid gas emissions? No, there are assuredly two effects, one of which is the addition of gases to the environment which may have a direct impact on human health (SO_2 and NO_2) or indirectly by formation of additional sulfate particulate in condensation nuclei. These have other indirect effects through particulate matter interactions in the lung. There are also some indications that cloud layers and fogs tend to concentrate acid gas droplets which may have a detrimental effect on human health and/or forests at high elevations where fog and cloud interactions are much more common.

With respect to the second impact, it appears that the highest correlation between forest damage and acid deposition from fogs or clouds is *altitude*. The decline of red spruce forests at high altitudes has been found to be correlated with greater percentages of the time when these regions were encircled by clouds and/or fogs. As noted in Chapter 2, direct plant impact in lab studies using acid mists has been noted due to leaching of nutrients from tree foliage and the plant crown. Thus, the more concentrated forms of acidic components in fogs or clouds appear to be better correlated to plant damage.

Natural wash out of soluble ions will occur at higher altitudes by gravity over the course of time. This leads to increasingly poorer nutrient loadings for vegetation at those elevations. The net effect is increasingly sparse soil with ongoing reductions in yearly growth, sap flows, and resin. Deficiency in essential proteins reduces a plant's ability to fight disease and to resist insects. The result is an apparent dying environment.

STRATOSPHERIC OZONE

There is probably no more greater uncertainty in issues affected by air pollution (apart from global climate change) than in the potential for anthropogenic air emissions impacts on the stratospheric ozone layer. These two areas are interrelated by virtue of the chemical substances which are involved. The reason for the concern for stratospheric ozone changes lies in the fact that certain simple molecules absorb incoming sunlight which contributes to decreases in surface radiation. As seen earlier, there are significant differences between the high altitude intensity of incoming solar radiation and the intensity measured at sea level.

While it is true that ozone at the earth's surface is a deleterious material, at very high altitudes it has a beneficial effect by blocking certain wavelengths of harmful solar radiation. In Figure 8.1a, we see a comparison of radiation curves over the entire wavelength of incident radiation, from 0.1 to 100 microns (on a logarithmic scale) and that generated by two surface temperatures: 5800° and 245° K. The curve to the left is the incident radiation from the sun's surface (at 5800°) as a function of wavelength. The curve to the right is the emission characteristic of the earth's surface temperature as a function of wavelength. Incident solar radiation appears in the higher energy (shorter) wavelengths, whereas that of the earth and its re-radiation is concentrated in the longer or *infrared* regions. The sun's radiation is concentrated in the *ultraviolet* (0.2 to 0.4 micron region) with a "window" in the visible light range of 0.4 to 0.7 microns.

The overall effect of altitude on radiation absorption is seen in Figure 8.1b. All of the gases from ground level to stratospheric levels contribute to light absorption (dark areas). A comparison of the radiation absorption at higher altitudes (11 kilometers) to that at the surface indicates the significant absorption due to the depth of the atmosphere and its constituent gases.

Figure 8.2 illustrates the *relative absorptivities* of different gases in the

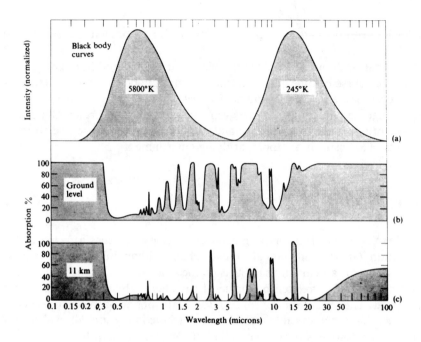

Figure 8.1.   Radiation emission and absorption curves.

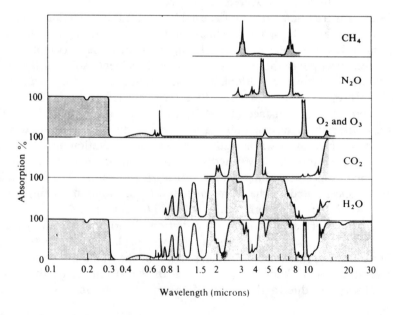

Figure 8.2.   Molecule-specific absorption curves.

top five bands, with a summary in the bottom band. All of the gases in these five bands are energy-absorbent gases and each has its own characteristic absorption wavelength. Water is the greatest *greenhouse* gas of all, due to its strong absorption from 0.8 to over 15 microns, which spans the infra-red region.

Of significant concern for stratospheric ozone is the third band in Figure 8.2. This shows the very strong absorption characteristics of oxygen and ozone in the ultraviolet regions between approximately 0.2 and 0.3 microns. Ozone is primarily responsible for this absorption. The absorption characteristics of other gases in the regions greater than 1 micron are the concern of global climate issues discussed later.

Stratospheric ozone serves as a protective layer for the surface of the earth, since it is known that ultraviolet radiation may have harmful effects not only on human health, such as skin cancer, but potentially on the phytoplankton in the earth's oceans as well.

Stratospheric Ozone Formation

The ozone in the upper atmosphere has always been assumed to be in a steady state condition. Equations 8-5 through 8-7 indicate the general chemical reactions occurring in the upper atmosphere.

$$O_2 + h\nu \rightarrow 2O^* \tag{8.5}$$

$$O^* + O_2 \rightarrow O_3 \tag{8.6}$$

$$O_3 + h\nu \rightarrow O_2 + O^* \tag{8.7}$$

These gases are apparently in a natural equilibrium, absorbing ultraviolet radiation to form ozone and then reforming oxygen with absorption of additional ultraviolet radiation. The formation and equilibrium concentrations found in various parts of the earth's atmosphere do vary according to latitude, wind velocity, sunspot activity, and temperature.

Early Observations

The reactions above were not of concern until recent observations apparently indicated a disturbance of the ozone/oxygen equilibrium in the Antarctic in the 1980s. It appeared that certain man-made chemicals had a correlation with decreases of the stratospheric ozone column during October (the southern hemisphere's springtime). It was postulated that anthropogenic emissions of chlorine-containing compounds, including certain refrigerant gases called chlorofluorocarbons, contributed to the perturbation of the stratospheric ozone equilibrium. Lab studies indicated they had a part in scavenging the ozone radicals which depressed the overall formation rate. The chlorofluorocarbons (CFCs) and related bromine-containing compounds (Halons) are used as refrigerants, solvents, fire-extinguishing agents, and in automobile air conditioning units, as well as industrial foam-blowing agents.

Regulations have been implemented which phase out chlorofluorocarbons in the United States by 1996. Other nations have agreed to phase out the use of chlorofluorocarbons by the year 2000. Since the diffusion rate of chlorofluorocarbons into the stratosphere is not instantaneous and there are latitudinal variations in concentration, there is a suggested lag time of 20 to 30 years in the maximum effect of chlorofluorocarbons to the destruction or depletion of ozone in the stratosphere. Therefore, incremental additional global ozone depletion in the Antarctic may be expected for some years to come, though the rate may be expected to diminish.

Natural sources such as volcanoes may have a significant impact quite apart from the impact of chlorofluorocarbons in terms of ozone-depletion potential. Scientists estimate that volcanoes typically dump 12 million tons of hydrochloric acid into the atmosphere annually, but only a portion reaches the stratosphere. The 1976 eruption of Mount St. Augustine (Alaska) deposited over 175,000 tons of chlorine compounds into the stratosphere. Some scientists recall that the 1982 eruption of El Chicon in Mexico thinned the ozone column by 20% as the chlorine-containing volcanic cloud mixed with the lower portions of the ozone layer.

There are historical disputes as to whether the ozone column changes noted in the last 15 to 20 years are truly a result of anthropogenic emissions. For instance, the amount of ozone depends directly on the flux of ultraviolet light from the sun, which varies with the 11-year solar cycle. There are shorter cyclic periods in solar output which will also change stratospheric ozone concentrations. Increases in sunspot activity, therefore, could be expected to, and do indeed contribute to higher ozone levels in the stratosphere. Satellite data show variations between 0.25% and 0.65% in the stratospheric ozone content every 13.5 days. These variations correspond to changes in ultraviolet emission from the sun, thus verifying that there are shorter time periods of solar output variability which also contribute to ozone variations.

Lab Studies

The reactions which are found to occur in the laboratory and appear to show correlations with ozone depletion in October in the Antarctic are summarized in Equations 8.8 through 8.10:

$$CF_2Cl_2 + h\nu \rightarrow Cl^* + CF_2Cl^* \tag{8.8}$$

$$Cl^* + O_3 \rightarrow ClO + O_2 \tag{8.9}$$

$$ClO + O^* \rightarrow Cl^* + O_2 \tag{8.10}$$

Lab studies indicated that chlorofluorocarbons, such as Freon 12 (seen above) slowly migrate to the upper atmosphere and also absorb incident radiation to provide a free radical chlorine atom plus free radical Freon fragments. In the next step, the free radical chlorine attacks ozone molecules to yield chlorine monoxide (ClO) and molecular oxygen. The chlorine monoxide further reacts by scavenging atomic oxygen (free radicals) to yield, once again, the free radical chlorine atom plus oxygen.

The overall effect, therefore, is for the chlorine free radical to destroy ozone as well as oxygen free radicals, which disrupts the normal equilibrium state. This was held to be responsible for diminishment of ozone concentrations in the stratosphere. The reason CFCs were found to be important is that they are virtually nonreactive in the troposphere and will slowly diffuse to the stratosphere where ultimately they will be exposed to high altitude ultraviolet radiation. That ultraviolet radiation is sufficient to split the molecule, yielding the free radical chlorine atoms.

Laboratory studies indicate that the ozone destruction effectiveness of a chlorine free radical is between 10,000 and 100,000 oxygen free radicals before it is ultimately removed from the process by reactions with hydrogen-containing molecules to yield HCl.

Antarctic Studies

The effect above was originally noticed as a short-term phenomenon in the extreme southern hemisphere in early spring, rather than a continuous depletion process. Some salient facts on the characteristics of the atmosphere over Antarctica are helpful in the attempt to understand these phenomena.

First, the air over Antarctica is isolated from the rest of the global circulation patterns during the winter. This is the result of the lack of air disturbances in the higher southern latitudes due to fewer continental land masses in that hemisphere. This leads to the formation of the isolated *polar vortex* in which the local atmosphere is cut off from other air currents. Also significant is the 10 to 20°K colder temperatures over Antarctica than over the Arctic. This condition causes stratospheric ice clouds, which are not seen to the same extent over the northern high latitudes.

The key appears to be the sudden release of reactive chlorine at the end of the Antarctic winter (September/October). Nitrogen dioxide reacts in the gas phase with chlorine monoxide to yield chlorine nitrate — $ClONO_2$. This compound can convert other chlorine species into chlorine gas which readily dissociates into chlorine free radicals. The chlorine mononitrate, *when condensed*, reacts rapidly with heterogeneous materials on the surface of an ice particle but not in the gas phase. Thus, at the very cold air temperatures above Antarctica, virtually nothing happens during the "winter" months, except to convert chlorine mononitrate into a solid form on an ice particle surface. With the first appearance of sunlight in the southern hemisphere springtime, chlorine free radicals are readily formed to enter into reactions with ozone, causing a drop in ozone concentrations during those weeks early in the springtime.

This sequence, coupled with the breakdown of the polar vortex and heating of the atmosphere in the springtime, causes mixing and dilution of the polar vortex gases with ozone-containing atmospheric parcels from the southern hemisphere to reestablish the ozone layer over the Antarctic. It has been observed that the gradients in concentration of ClO and chlorine mononitrate appear to shift by as much as 5° in latitude from one day to the next. This illustrates the importance of the disturbances of the polar vortex in determining the chemical compositions of the Antarctic air.

With respect to the northern hemisphere, these effects are not seen due to the warmer temperatures, better atmospheric mixing, and lack of available ice particles to create the sudden loss of stratospheric ozone.

UV Data and Other Impacts

One of the major concerns for potential stratospheric ozone depletion is that UV radiation will increase. During the period when CFC concentrations in the stratosphere have been increasing, the ground level UV radiation has been monitored. Should the theory of ozone depletion be correct, it would be seen in these measurements, unless other natural phenomena are of greater impact.

From these monitored data it has been found that ultraviolet radiation over the United States *has decreased*. Figure 8.3 shows the measurements of

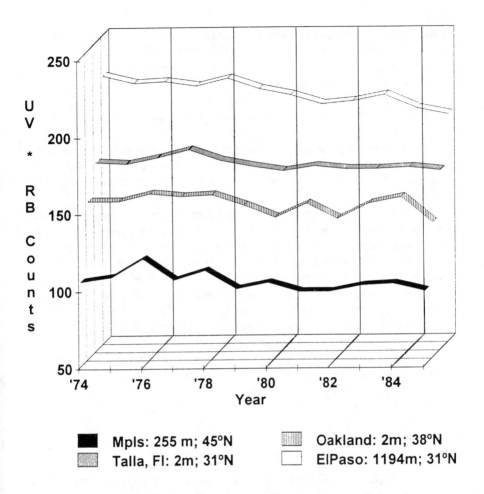

Figure 8.3.   U.S. ground level UV radiation, 1974 to 1985.

decreasing UV radiation of as much as 7%. In fact, it has been found that ultraviolet radiation over the United States *has decreased* during the monitored period 1974 through 1985.

These sites illustrate the natural variation in UV radiation to be found at various locations representative of different latitudes and elevations. Tallahassee (FL) and Oakland (CA) are at sea level, Minneapolis (MN) is at 255 meters above sea level, and El Paso (TX) is at 1194 meters. El Paso and

Tallahassee are at about the same latitude (approx. 31° north). Oakland is at about 38° and Minneapolis is at about 45° north latitude.

General trends can be seen in this monitored data. From the figure it appears that the higher the *elevation* (El Paso vs. Tallahassee), even at about the same latitude, the higher the UV radiation. Also, the higher the *latitude* (all sites), the *lower* the incident UV radiation. Location, thus, appears to be the most significant factor.

Other natural influences which have been proposed include sunspot activity, atmospheric turbidity, humidity, and cloud cover as factors in explaining these variations. Nevertheless, while CFCs have been increasing, UV radiation has decreased at all sites.

In other studies, National Oceanic and Atmospheric Administration scientists indicated that in the mid latitudes of the northern hemisphere, rural ultraviolet radiation *declined* between 5% and 18% during this century. At the same time, the *Journal of Physical Research* (September 20, 1991) uses NASA data to demonstrate that overall global ozone levels have *increased* in recent years at an average rate of approximately 0.28% per year. Thus, pure measurements of ozone concentrations in the Antarctic may not be giving a clear picture of global ozone trends. Likewise, average effects and measurements for different locations may yield significantly different patterns, or patterns with no statistical significance in certain areas. This is one of the paradoxes still unresolved to the satisfaction of all.

Alternatives

With the phaseout of chlorofluorocarbons, the challenge has become one of finding substitutes which may be used to provide for refrigeration and air conditioning. To a large degree, uses of CFCs and other chlorine-containing organics for solvents and foam-blowing agents have been replaced by alternatives, some of which are water-based systems.

Hydrochlorofluorocarbons (HCFCs) will probably be the interim refrigerant gases while CFC designed and containing equipment nears the end of its useful life. Chemicals which are nontoxic, nonflammable, and have lesser potentials for ozone depletion will be the chemicals of choice.

Ultimately, substitute refrigerant gases such as the HFCs (hydrofluorocarbons) may be candidates since they have a zero ozone-depleting potential. However, they have a higher energy penalty compared to CFCs. These alternative gases will require changes in the design of the equipment to provide equivalent amounts of refrigeration.

GLOBAL CLIMATE CHANGE

One of today's current concerns is the suspected impact of anthropogenic air pollutant gases on global climate. Other concerns are for potential secondary impacts such as rises in sea level, flooding of low-lying wetlands, impacts on agriculture, etc. if there is a proven long-term atmospheric temperature rise or *greenhouse effect.*

To address this, we must first ask, is there a *greenhouse effect* which causes a warming of the earth's surface? The answer is yes, for without a greenhouse effect due to gases in the atmosphere, the average temperature of the earth's surface would be $0°F$. As a consequence of greenhouse gases, the average temperature of the earth's surface is approximately $60°F$. Thus, we do experience a global warming impact as a result of greenhouse gases. The next question is, does the scientific data show an effect due to anthropogenic gases? Also, how much of the estimated contribution to global climate change do these gases make, and finally, are all of the effects "bad"? We will investigate some of these items in the following sections.

Most of the impacts attributed to anthropogenic gas emissions are based on weather models attempting to predict temperatures and temperature effects. The difficulty lies in the fact that weather forecasting computers cannot truly project what any given location's temperatures will be within the next week or month, much less decades in the future. Thus, projections for temperatures in the next century are even less sure, particularly as a result of air pollutant-generated global climate change.

The controversy is in attempting to verify trends in small temperature changes (less than $1°C$) in the face of day to night fluctuations as high as $15°C$. Summer to winter variations may be as high as $30°C$ or more. The other problem lies with *perception*, i.e., if it is snowing in Jerusalem and New York City is having its hottest summer on record, it is hard not to believe that we are facing severe climate changes. As Steven Schneider of the National Center for Atmospheric Research was quoted, "It's possible that everything in the last 30 years of temperature records is no more than noise."

Historical Perspective

Direct measurements of the factors influencing our global climate over time are restricted to the immediate past decades. Precise measurements of air temperatures and surface sea temperatures have been made for only a few hundred years. Accurate measurements of certain atmospheric parameters, such as stratospheric ozone and chlorofluorocarbons concentrations, go back less than 40 years. The first recorded quantitative measurements of methane in the atmosphere were only made in 1948. Atmospheric measurements of methane were still being correlated in the 1980s.

Consequently, a consideration of historical records is in order since, presumably, climate changes in the past occurred without the influence of significant anthropogenic air emissions.

It is known that prior to about 1800 BC, the entire Middle East experienced a cool, wet climate and forested terrain. By about the 1st century BC, North Africa was functioning as the grain-growing "breadbasket" for the Roman Empire. Thus, the implication that this geographic area experienced a much cooler, wetter climate over a time period of over 1000 years is considered valid. That this did occur has been verified by satellite photos which are able to penetrate the surface sands of North Africa. These surface penetrating photos show widespread riverine courses beneath the sands, which indicate much more extensive runoff and ongoing wetter climates in previous millennia.

In the 10th to 13th centuries A.D., Europe experienced the Medieval Warm Epoch or "Little Optimum," in which for several hundred years, the climate was milder, warmer, and much more conducive to society in general. Wine grapes were grown in England, which is not possible today. At the beginning of the 14th century, however, dramatic climate changes occurred worldwide.

The Baltic Sea froze over in the years 1303 and 1306–1307. Beginning in 1310, cold storms and rises in the Caspian Sea levels were noted throughout the balance of that century. In 1350, in northwest India, the river Mihran disappeared due to extensive droughts. Monsoons in that era were so weak that widespread starvation due to droughts also occurred. In 1372, Chinese records indicate that sunspot activity increased to its maximum over a two-century period.

In that century, the Viking settlement of southern Greenland provided a well-attested fact of the influence of climate change and its historical effects. Having been settled in the 10th century during the "Little Optimum," the climate turned severe with the start of the 14th century. The sites of old farms, now buried in ice, indicate the extent of their colonial effort, and the graves of the farmer settlers contain thick masses of plant roots which must have formerly grown under less severe temperatures than those of today. These are now frozen in soil.

This lower temperature effect is further verified by records indicating that marauding bands of Eskimos, who lived by hunting seals *near the edge* of the Greenland ice pack, attacked and destroyed the farming settlements in southerly movements throughout the 14th century. Their southerly movements and records of attacks indicate that the ice was spreading southward during that era.

This was the beginning of the "Little Ice Age" that lasted for hundreds of years and was virtually a global event which reached its maximum between the mid 16th and mid 19th centuries.

From approximately 1850, however, at the tail end of the Little Ice Age and on into the 20th century, a warming trend has been experienced. Indeed, certain glaciers have been found to be retreating in Europe. But the glaciers in central Greenland have been found to be growing in the earlier half of the 20th century.

Figure 8.4 is an illustration of the known and estimated temperatures in the Great Britain area for approximately the last 10,000 years compared against the average for the first half of the 20th century. Approximate

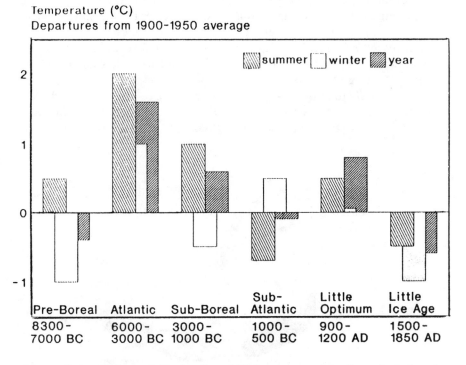

Figure 8.4.   Estimated England and Wales temperature for selected periods during the last 10,000 years.

departures from the 20th century average temperature are shown for summer, winter, and annual averages.

Over all of these historical events have been the geogenic factors, such as volcanoes, which are proven to dramatically affect climate, worldwide, for several years following large volcanic eruptions. The 1815 eruption of Mt. Tambora in Indonesia, which gave rise to the "year without summer" in 1816 in America, is an example of other effects which may certainly impact climate. The eruption of Mt. Pinatubo in 1991 depressed worldwide temperatures a full degree or more for two years following that event.

With respect to the 20th century, a great deal of controversy continues.

Current Concerns

Concerns are usually expressed for emissions of greenhouse gases which contribute to absorption of infrared radiation. The greenhouse effect leads to rises ultimately in atmospheric temperature. The greenhouse gases are seen in Figure 8.5 by their approximate contributions to atmospheric infrared radiation absorption at current average concentrations.

The *major infrared-absorbing gas is water*. Therefore, discussions which ignore water vapor do so at the risk of losing a true understanding of the causes of the greenhouse effect. Most popular estimates of greenhouse gases refer *only* to anthropogenic sources of these respective gases. This is of significant concern, since human beings contribute only 5% to the total CO_2 emissions to the atmosphere on a global budget basis. Water vapor, at

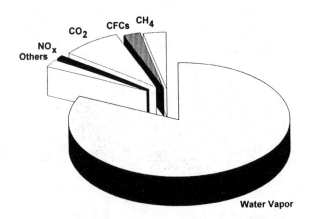

Figure 8.5. Greenhouse gas contributions.

Table 8.4. Global Methane Emissions

Source	10^6 TPY	%
1. Swamps, lakes and marshes	146.6	28.8
2. Rice paddies	142	28
3. Animals	72	14
4. Anthropogenic	62	12
5. Jungles and forests	60	12
6. Biomass burning (all)	25	5

TPY = tons per year
Source: Dr. Don Blake, Ph.D. Dissertation, University of California, Irvine, 1984.

0.01 to 7% of the total atmosphere, is the largest single contributor, and is not counted in most global climate change scenarios.

Most of the interest to date has focused on anthropogenic emission sources and the rise in atmospheric CO_2 levels. The problem is that when we compute global budgets for CO_2 emissions, compare them to measured CO_2 concentrations and adjust them for known sources and sinks, at least 40% of the total cannot be accounted for.

Methane is also a greenhouse gas and its concentration is slowly rising. Global methane budgets, as seen in Table 8.4, indicate that anthropogenic source emissions of methane are approximately 12% of the total, with over half coming from swamps, marshes, lakes, and rice paddies. Thus, the anthropogenic contribution is only a small player in the methane scenario. There is also a gradient of methane as a function of latitude on the earth's surface, as indicated in Figure 8.6, which shows higher concentrations of methane in the northern hemisphere decreasing to lower concentrations in the southern hemisphere. This indicates that methane emissions are associated primarily with land masses in the northern hemisphere.

The cause of concern for chlorofluorocarbons as a greenhouse gas is that they are approximately 10,000 times more effective at absorbing radiation than is CO_2. Nitrous oxide (N_2O) has strong absorption in the infrared radiation band. However, as seen earlier, N_2O is the primary NO_x species given off naturally. Indeed, N_2O is ubiquitous in the atmosphere and occurs at concentrations of approximately 330 ppb worldwide.

The biggest single problem is that we have very little control over the majority of the emissions of carbon dioxide and practically none over water vapor, nitrous oxide, and methane. Thus, our approaches on a global basis must take these facts into account in our air quality management strategies.

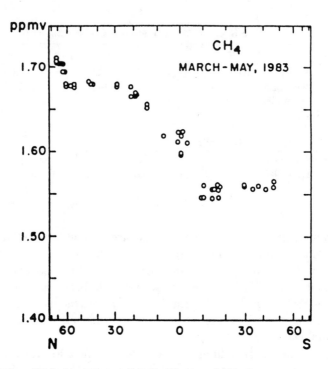

Figure 8.6.   Global methane distribution vs. latitude.

Known and Measured Data

Direct measurements are best for attempting to make projections based upon short-term fluctuations in emissions. For instance, the eruption of Mt. Pinatubo in June 1991 cooled the earth's average surface temperatures by 1–2°F, which overwhelmed projections based upon the anthropogenic gases in the atmosphere. Thus, short-term natural phenomena may cause significant variations which cannot be accounted for in models.

Most of the recent concern for carbon dioxide has been generated by reviewing CO_2 data over the last 30 years. Typical of these measurements are those at the Mauna Loa observatory in Hawaii. In Figure 8.7 are plotted the atmospheric CO_2 levels at this site over the last 30-plus years. There has been a decided upward concentration trend over that period. Attempts at correlations between gas concentrations and measured average temperatures are much more difficult. The difficulty arises in that observations of temperatures at the earth's surface do not track the same rises in CO_2 concentrations. In Figure 8.8 are the temperatures over the last 130 years in

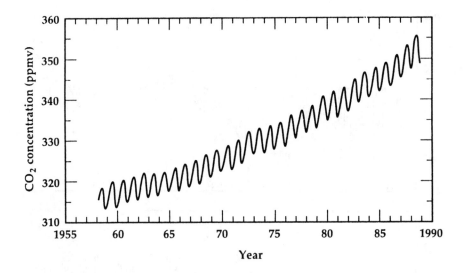

Figure 8.7. **The rise in atmospheric carbon dioxide.**

the northern and southern hemispheres, which indicate that the majority of the temperature increase (approximately 0.7°C to 0.8°C) occurred prior to 1940. This indicates little correlation of temperature increase in the northern hemisphere with the rises in CO_2 noted in Hawaii. Closer inspection of these figures indicates greater variability from year to year than in average trends. Both hemispheres have noted increases over this period. Other problems, of course, are that even with the best temperature records, the numbers are "fuzzy" by several tenths of a degree, either plus or minus.

On a regional basis, i.e., the contiguous United States, Figure 8.9 shows the observed temperatures for approximately the last 100 years in the continental United States. This indicates virtually no change in average temperatures for the past century when corrected for the urban heat island effect, as noted earlier.

When one is looking at a shorter time frame, *the uncertainty becomes greater.* By taking smaller increments of years, greater fluctuations are seen and pattern recognition becomes even more difficult. While the north hemisphere had a slight temperature increase over the last 50 years (which corresponds to the greatest increases of anthropogenic gas emissions, such as CO_2 and CFCs), the United States experienced a slight cooling trend. Of course, the year-to-year variability is much greater than the calculated variation of temperatures over a mean value.

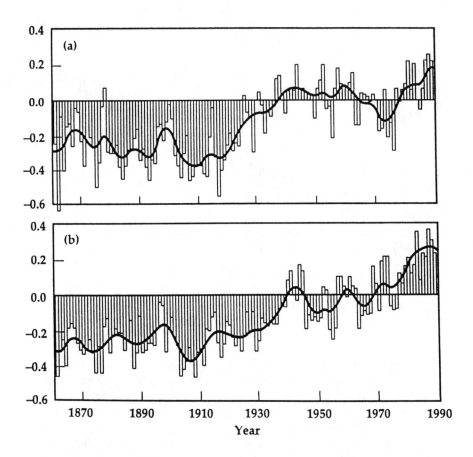

Figure 8.8.   Observations of mean temperature in the Northern and Southern Hemispheres.

Other trends in maximum daily and minimum nighttime temperatures in the contiguous United States over the last 100 years indicate that the maximum daily temperatures increased up until about 1935, and since then have been decreasing. At the same time, the minimum temperature has been rising. Thus, if a short period trend analysis conclusion were drawn, it would be that both the daytime high temperature and nighttime low temperatures were approaching an equilibrium value, and the overall climate was becoming more moderate.

Apart from temperatures, a look at the changes in sea level on a global

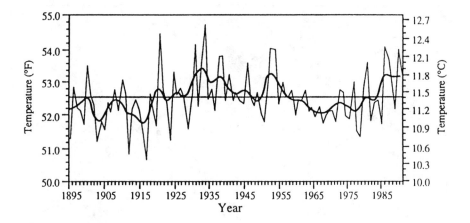

Figure 8.9.   Annual average temperature for the contiguous U.S. from 1895 to 1990.

basis yields a different pattern. If global warming is truly occurring, a clear pattern of rising sea levels would be expected. Figure 8.10 is an illustration of changes in worldwide sea levels between 1960 and 1979 around the world. What has been observed is that the sea level is rising in some locations and falling in others. Indeed, at some locations in close proximity, such as southern West Africa, both sea level rises and falls have been observed. It is therefore difficult to say that any overall sea level rises have in truth occurred.

Natural Variability

Changes in sunspot activity have been alluded to with respect to potential climate change on an historical basis. The sun's brightness, a measure of its radiation intensity, affects global temperatures. Likewise, changes in the magnetic field flux of the sun could impact the earth's atmosphere and therefore the global heat balance. Indeed, these may be the events resulting in natural climatic variability. Changes in solar brightness or magnitude do occur, as seen in Figure 8.11. The changes are for two stars with nearly identical properties to those of the sun over a four-year period. Both of these stars indicate a change in brightness over very short periods of time. Likewise, changes in magnetic surface activity in one of those stars is seen in

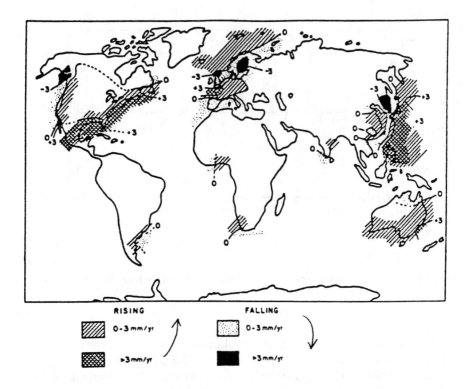

Figure 8.10. **Average annual rates of change in worldwide sea level between 1960 and 1979, as determined from data supplied by the Permanent Service for Mean Sea Level, Bidston Observatory, UK.**

Figure 8.12. This shows a cycle of approximately eight years, which is a fairly short fluctuation in stellar activity. Such variations in the sun could also induce temperature fluctuations in the earth's atmosphere.

As noted earlier, the measured changes in air temperature do not match the emission rates of man-made greenhouse gases. However, it appears that the pattern of temperature change does match (correlation coefficient of 0.95) the pattern of changes in our sun's solar activity for over the last 125 years. Figure 8.13 is a comparison of solar activity and average global temperatures from 1860 to 1985. When solar activity increased in the first half of the 20th century, global temperatures also increased. Likewise, temperatures declined when solar activity declined. Solar activity is measured by cycle length.

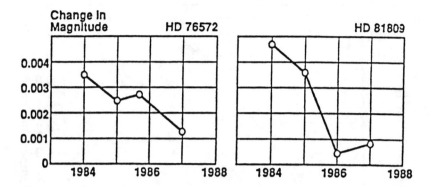

Figure 8.11.   Changes of 0.23% (left) and 0.42% (right) in the brightness of two stars with nearly identical properties to those of the sun, over a four-year period.

Other Considerations

A retrospective analysis of historical records and other evidence indicates that significant climate changes are caused by changes in the sun's brightness over long-term time scales (centuries). An analysis of climate records for England (alluded to earlier) indicates that during the last 1000 years,

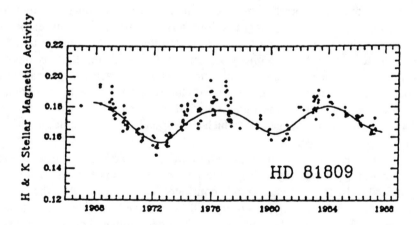

Figure 8.12.   An 8-year cycle of changes in magnetic surface activity in HD 81809.

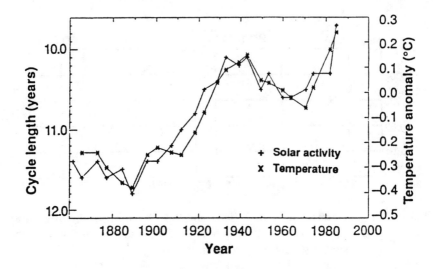

Figure 8.13. Comparison between global temperatures (x) and solar activity, measured by the length of the solar cycle(+).

protracted periods of cold weather occurred roughly every 200 years. As noted, this includes the Little Ice Age where the average temperatures in Europe were roughly 1°C lower than they are today. The decade of the 1690s was the coldest on record in the history of Europe.

There have been correlations of solar activity with carbon-14 content of tree rings. These yield an integrated long-term record of apparent temperature changes corresponding to carbon-14 content. Comparing carbon-14 records to climate history, one group of researchers found that six of the last seven most severe decreases in solar activity (as measured by carbon-14 content in the tree rings) "corresponded closely to cold spells in the climate record. One of these is the famous period of about 50 years of very low solar activity in the 17th century which corresponded with the coldest period of the Little Ice Age."

Sunspot records confirm that a minimum of solar activity occurred in the 19th century followed by a rise in the 20th century, as noted earlier, to the current levels. If past cyclic trends continue, we can expect that the rising solar activity of the current century will be followed by a decline of solar sunspot activity in the 21st century. This decline would lead not to a temperature rise, but to a temperature cooling as a result of natural forces.

Feedbacks

Feedbacks are those natural phenomena which act to oppose wild swings in temperature in the atmosphere. One feedback which moderates global temperature changes are changes in the water content of the atmosphere. Any phenomenon which increases air temperatures will increase moisture due to greater driving forces for evaporation over the ocean. When this occurs, warm moist air will rise until it meets cooler temperatures at higher altitudes, which will give rise to clouds. Clouds, due to their high reflectivities, act in the opposite direction to a global warming scenario. Any increases in cloud cover above normal would increase the reflectivity of the earth. This would cause cooler temperatures in the lower regions of the atmosphere, thus acting as a natural "brake" to anticipated global warming scenarios.

Other researchers have found unambiguous evidence that the presence of clouds reduces the earth's mean surface air temperature. In fact, they found that the size of the observed net cloud forcing (climate change) is about four times as large as the expected value of radiative forcing from a doubling of CO_2 concentrations. Increasing clouds act in the opposite direction to increasing CO_2. Thus an increase in planetary cloud cover would offset much of the effect from doubling or even tripling the earth's CO_2 concentration.

Models

Expensive attempts have been made to mathematically model all of the physical, meteorological, and other impacts which may lead to overall global climate change. These models are called general circulation models (GCMs) and are designed to simulate temperature impacts due to the physics of the atmosphere. The difficulty with global climate models is that they only grossly approximate conditions known to be occurring, and they only handle large (continental size) land masses. Unfortunately, the biosphere is so vast and complex that even advanced computers can only deal with approximations.

Due to these limitations, the GCMs have serious defects, notably because they fail to predict the present based on historical input data. That is, when they are run in reverse, they are not able to duplicate what is known to have occurred. For instance, researchers have set global models to the conditions of 1880 and found that the models indicate a rise in global temperatures by as much as five degrees. However, the actual increases in the last century are, at best, only one degree. Thus, climate modelers are forced to "re-tune" or adjust their models to calculate projected warming trends. At best, these models may come only within two degrees of what has actually occurred.

Figure 8.14 indicates the results of one such model which was set to start in 1880 and ran through the mid 1980s. The problem is that the true temperatures did not match the observed temperatures except at the initial and final points (which were set when the model was begun).

Another major problem is in dealing with water in the atmosphere. Early models assumed that water behaved the same as either vapor or as ice crystals. When a British meteorological office adjusted their global climate models to take into account actual experiments (which indicated large differences between the two states of water), the predicted greenhouse warming fell from about ten degrees to approximately three degrees.

Other difficulties which at the present time are not factored into the general models are the impacts of contaminants such as sulfur dioxide.

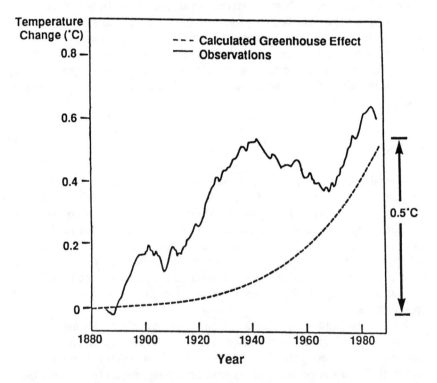

Figure 8.14.   Calculated warming due to the increase in greenhouse gases in the last 100 years (dashed line), compared with observed temperatures (solid line).

Sulfur dioxide forms particulate sulfate and sulfuric acid aerosols, which have been found to act as additional radiation reflectors. Sulfates and aerosols will lower overall temperatures if found in the atmosphere. Thus, one control approach, i.e., reducing sulfur dioxide, may contribute to global warming (if it is occurring) by removing the additional reflectivity noted here.

With respect to sea level changes, the predicted rises have gotten smaller over the last 15 years. In Figure 8.15 are the results of modeled greenhouse effects and the resultant predictions for sea level rise. Original attempts modeled sea level rises at approximately 25 feet, whereas today, these predicted sea level rises may *approach* one foot at best, which is *less than the natural daily variation* in tides.

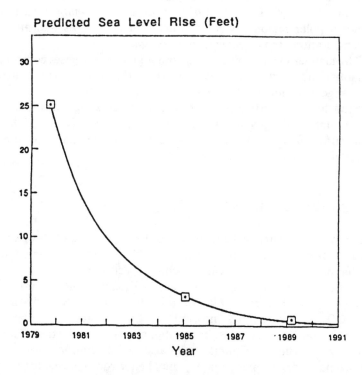

Figure 8.15.   **Predictions of the rise in sea level resulting from the greenhouse effect, plotted against the year in which the prediction was made. The predictions decreased from 25 feet in 1980 to three feet in 1985 and to one foot in 1989. Today predictions are given in centimeters.**

Recent Research Findings

A consortium of researchers from around the world formed the Model Evaluation Consortium for Climate Assessment (MECCA). This group has attempted to evaluate modeling results for a variety of greenhouse gas concerns through the present and balance them with the levels of uncertainties mentioned here. Some of their key findings to date are:

1. As CO_2 increases, the *rate of global warming decreases*. Increases in CO_2 in the atmosphere produce a less than linear increase in mean temperatures in the mid latitudes. This "leveling off" effect on temperatures appears to be more related to saturation of the model's water vapor absorption than of the CO_2 absorption. This indicates the key importance of water in global climate change.
2. Thirty years of observation *did not detect a warming trend* in the north polar regions, which is where current GCMs indicate that temperatures *should have risen even more*.
3. The manner in which each of the other greenhouse gases control the global radiation balance is different from CO_2. This alone will change assessments of global warming potentials.
4. Modeling on regional scales is critical for creating realistic understandings of the greenhouse gas issue. To provide this level of detail, a tenfold increase in currently used computing capacity would be required.

ALTERNATIVE VIEWS

A number of researchers have indicated that increasing levels of CO_2 and moisture content would have positive benefits, even if accompanied by slight temperature rises, particularly with respect to agriculture.

For instance, a doubling of the carbon dioxide concentration in the atmosphere would increase plant productivity by almost one-third. In higher CO_2 concentrations, most plants grow faster and bigger with increases in leaf size, thickness, branching, and seed production. The number and size of fruit and flowers would also rise. In addition, root and top ratios would increase, giving many plants better root systems for access to water and nutrients.

The effects from higher CO_2 are due to: (1) a superior efficiency of photosynthesis; and (2) a sharp reduction in water loss per unit leaf area. A related benefit comes from the partial closing of pores and leaves associated with higher CO_2 levels. Closing of the pores is a major source of moisture retention. Thus by closing them, higher CO_2 levels will greatly reduce plants' water loss, which may be a significant benefit in more arid climates. A major benefit of closing stomata cells due to higher CO_2 concentrations would be *less damage due to air pollutants.*

Increased Yields

This effect has been proven in greenhouse studies, including tomatoes, cucumbers, and lettuce. These show earlier maturity and *larger fruit size, greater numbers* of fruit, a *reduction in cropping time,* and *yield increases averaging 20 to 50%.*

Cereal grains show yield increases from 25 to 64% with corn, sorghum, and millet. Sugar cane showed increases from 10% to 55% (resulting primarily from superior efficiency of water usage).

Figure 8.16 shows the *rise in photosynthesis activity* for a rise in carbon dioxide concentrations from 340 to 640 ppm, thus illustrating the effect noted earlier. A rise in temperature would also increase net photosynthesis, as seen in Figure 8.17 for three other crops.

Trees, which cover approximately one-third of the earth's land area, account for two-thirds of global photosynthesis. They likewise respond favorably to higher concentrations of CO_2. Fruit tree production is verifiably enhanced. When concentrations of CO_2 are increased by a factor of two, orange trees yield 2.8 times more biomass in five years, and ten times more oranges in their first two years of production.

Other Effects

Rising CO_2 levels also compensate for lower light intensities in the high latitudes. Flowers and vegetables grown in CO_2-enriched greenhouses experience higher percentage boosts in plant productivity under very low light intensities than when under normal light conditions. Enrichment of the air by CO_2 also appears to offer some protection to plants against both extremely hot and cold temperatures.

There is also evidence that increased levels would raise the optimal temperature for plant growth. If significant global warming occurs, and if the higher CO_2 world of the future leads to higher temperatures, plants could be expected to respond favorably to increases in both CO_2 and temperature. Increases in CO_2 and in water retention would provide *stress relief* to plants and *less ethylene gas (and photochemical ozone) production.* Fewer diseases due to stomata closure could be expected.

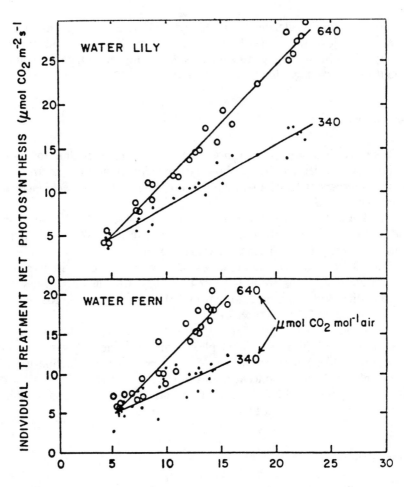

Figure 8.16.   Mean net photosynthesis (μmol CO_2 m^{-2}s^{-1}).

Secondary effects of higher CO_2 are that as crop yields rise with CO_2, the amount of land devoted to agriculture could decline. This would allow for *fewer fertilizers and pesticides* to be used for crop production. Plants, such as trees, which use a C_3 metabolism (so called because the first photosynthesis step produces three carbon atoms per molecule) benefit more in high CO_2 atmospheres than C_4 metabolism plants, hence a more rapid reforestation and an expansion in forest biomass could be expected. Of the 21 most important food crops, 17 have C_3 pathways. These include rice, wheat,

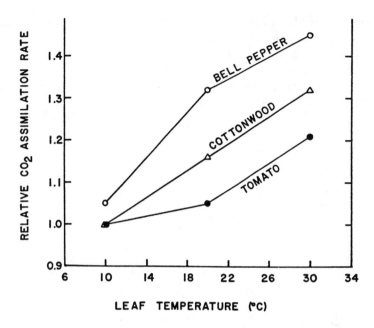

Figure 8.17. Relative CO_2 assimilation rate (net photosynthesis at 600 μmol CO_2 per mol air divided by net photosynthesis at 300 μmol CO_2 per mol air) as a function of leaf temperature.

barley, oats, soy beans, potatoes, sweet potatoes, casabas, sugar beets, and coconuts.

On the other hand, 14 of the 18 most noxious weeds are C_4 metabolic pathway plants, which are not as favored in a high CO_2 atmosphere. Thus, rising levels of atmospheric CO_2 would generally favor crop production over weeds.

A boost in plant production due to major increases in CO_2 would provide more food for birds, fish, and mammal populations as well. Also listed as benefits would be *soil stabilization, lowered soil erosion, and greater subsoil productivity*. These factors would provide more organic matter to the soil matrix, which would provide for soil enrichment, increases in the microbial population, increases in earthworms, and increases in salt tolerance.

Decreases in insect foraging would occur due to lower nitrogen to carbon ratios. This would decrease the nutritive value somewhat and therefore *fewer insects* would result. Insects would also decrease generally due to increased mortality and longer larval development times observed when such insects had been fed CO_2-enriched foliage. Lab tests at 650 ppm CO_2 found reduced populations of leaf-hoppers, predaceous flies, and pink boll worms on crops grown under those conditions.

Based on the above, the potential positive effects of a CO_2 increase should perhaps be considered prior to attempts to substantially decrease those concentrations as an air quality management strategy.

9 AIR QUALITY LAWS AND REGULATIONS

The use of the sea and air is common to all; neither can a title to the ocean belong to any people or private persons, forasmuch as neither nature nor public use and custom permit any possession thereof.

Elizabeth I, to the Spanish Ambassador, 1580

GENERAL APPROACHES

Laws and regulations are at the heart of our air quality management strategies, even as health and environmental effects are the justification for those laws and regulations.

Historically, a general law approach to air quality concerns has been taken. Outright prohibitions against certain activities generating air contaminants have been adopted for centuries. These have existed as early as the 13th century (Chapter 1). These prohibitions were for actions taken which harmed, or potentially harmed, the health or safety of the citizenry.

A second approach has been to allow for lawsuits as a result of public nuisance complaints. In general, public nuisance complaints have been among the oldest legal approaches used to abate air pollution emissions. Today, these are still used when odors or emissions occur which may affect more than one person. Public nuisance provisions have been used in some areas as one approach to enforcing excess cancer risks from noncriteria pollutants.

Other Approaches

Other approaches which are used to lower air emissions, and thus enhance the general air quality, include taxation, land-use controls, source specific emission standards, and standards based upon health risk. Taxes do not abate or reduce emissions in and of themselves. They form an indirect (economic) approach where, if the tax is too high, source owners will voluntarily reduce emissions to avoid paying them. This system is currently in effect in the United States to encourage reducing emissions of certain chlorofluorocarbons.

The control of land use, typically administered at the county or municipal level, attempts to separate sources of air contaminants from receptors (people) by sufficient distance so that natural air dispersion and dilution will lower contaminant concentrations to levels which will not generate a public nuisance. It also attempts to control growth and thereby limit emissions creation.

The backbone of current air quality management strategies has been the so-called "command and control" approach. In this approach, specific mass emission rate limitations are placed on sources. Receptors experience lowered concentrations of contaminants and therefore, better air quality.

Indirect source controls attempt to change societal patterns or personal behavior (driving gasoline automobiles with only one occupant) which indirectly contributes to air emissions. This is generally accomplished by requiring employers to institute incentives for employees to use alternate transportation, carpool, shift work hours, "tele-communicate," or to stagger work days. These are assumed to work by limiting the vehicle miles traveled (VMT) in an area and thus to lower mobile source emissions.

The most recent approach has been to evaluate health risks and to set, by regulation, acceptable levels of health risk for the receptors. Sources are required to implement whatever controls or limitations on emissions are necessary in order to avoid exceeding established excess health risks. These health risk-based approaches are primarily directed toward the noncriteria air pollutants.

The Process of Regulation

Within the United States federal system of laws and regulations with respect to air quality, there are specific procedures which must be followed.

The initial step, of course, is the establishment of the laws governing air quality. In this system, elected officials adopt legislation, and upon signature by the Chief Executive, the law takes effect. The laws set direction and goals, and identify those branches of government responsible for implementation. The agencies involved are mandated to formulate specific regulations which implement the goals, outlines, and intent of the law.

The process of implementing regulations is one in which the public has an opportunity to review, comment, and influence such regulations. In general, regulations identify the specific mandates of law and the particular problems and sources of those problems, such as air pollution. The problem could be nonattainment of a health-based air quality standard.

Role of the Public in Rule Making

Draft regulations proposed by government agencies are published in the *Federal Register*. This is the federal government's daily newspaper of all actions and activities of a regulatory nature along with background information, sources, etc. Draft regulations set time frames and dates for public hearings and written comment on such regulations.

Within the federal government, this is a three-step process in which the proposed regulations undergo public review in three stages. These are the pre-rule, proposed rule, and final rule stages. At each stage, the proposed regulation outlines its significance and legal authority, mandated deadlines, abstracts of the regulations, entities affected, and the responsible agency. At each stage, the public is invited to comment on the draft or re-drafted regulations. Once a final rule has been published, a deadline for a final public hearing is set, at which time the last public comments are received. Following the close of public comment, the agency *promulgates* the new regulation in the *Federal Register*. At that point, the regulation is final and takes effect.

Levels of Authority

Historically, the levels at which air quality regulatory authority existed has varied. In some municipalities, attempts at air pollution control under local health departments were in effect prior to World War II. The immediate post-World War II era saw a rapid increase in the number of regional and statewide air pollution authorities with varying degrees of responsibility. With the establishment of the Environmental Protection Agency in 1970, and the passage of the first comprehensive federal clean air act, the focus of air quality control authorities has been increasingly federalized.

The Environmental Protection Agency has nationwide authority in all areas of air quality management. In addition to setting air quality standards and maximum levels of emissions, the federal government has been extensively involved in monitoring, research, and funding local programs.

Each state has specific authority to implement its own air quality management program. Enabling legislation at the state level is required in order to set up a statewide program. To enforce federal laws, state legislation must be adopted and approved by the EPA to give federal enforcement authority to the state. Otherwise, the federal government is the enforcing authority.

With respect to specific provisions, states must be at least as stringent, or in cases of a waiver, may be more stringent in their regulations than the equivalent federal regulations and laws. There are some limitations. This occurs when there is a federal preemption or where there may be a conflict with the constitution, such as with the interstate commerce clause.

States may also adopt their own ambient air quality standards and provide for their own implementation plans. They possess independent legal authority to manage air quality within their jurisdiction. For example, California has its own Clean Air Act in the state's Health and Safety Code. In certain jurisdictions, there is no permitting authority at the state level. This may be delegated to local authorities.

In addition to the State Implementation Plan (SIP), local authorities in areas not in "attainment" are required to establish air quality management plans (AQMPs) which detail those activities and regulations which will be demonstrating reasonable further progress toward attaining the ambient air quality goals. Reasonable further progress (RFP) regulations are a part of the state implementation plan and are legally enforceable as a part of the SIP at the federal level, once they have been approved by the EPA.

In some areas, the regional or local air pollution control or management authorities were the first government agencies regulating air pollution in their areas. As a consequence, many have a long history of air monitoring, regulations, emissions standards, local enforcement, and permitting. In general, the regional or local air pollution authorities carry the burden of day-to-day activities with respect to implementation of air quality legislative mandates and regulatory requirements of both federal and state agencies.

Municipalities may also implement their own ordinances governing emissions of air contaminants within their jurisdiction, provided they are not preempted by or in conflict with other levels of government. For example, certain cities adopted local ordinances which banned emissions or use of chlorofluorocarbons (CFCs). Thus, every level of government may be involved in air quality management to a greater or lesser degree.

FEDERAL LAWS IMPACTING AIR QUALITY MANAGEMENT

There are a number of laws which impact our approaches to air quality management. They include those which are media-specific (water, solid waste), or which have an air quality component (toxic substances, nuclear materials). The former deal with some other aspect of environmental contamination but which have some impact on air quality. The latter includes the Clean Air Act and its predecessors.

Pre-1990 Air Quality Acts and Impacts

Prior to the formation of the EPA in 1970, a number of federal laws dealt with air quality. These primarily dealt with the criteria pollutants but did show some attempts to address noncriteria, air contaminant issues. A number of these major concepts were modified and incorporated into the 1990 amendments to the Clean Air Act.

The primary national ambient air quality standards (NAAQS) for the six criteria pollutants are the *driving force* for federal regulatory action due to their known health effects. The NAAQSs are periodically reviewed and subject to change as more information becomes available.

Implementation Plans

The concept of a *State Implementation Plan* (or SIP) with federal oversight is the basis of approach to air quality management. SIPs are ongoing documents which provide a regulatory framework for each state to demonstrate to the federal government that they are on a path to *attaining and maintaining* the national ambient air quality standards. Plans for the states which are not in attainment with those standards form a significant portion of each SIP. Federal law does *provide for preparation of a federal implementation plan* (FIP) by the EPA if state implementation plans are not approved or are deficient.

Monitoring and Limiting Emissions

Monitoring ambient air quality and limiting emissions of criteria pollutants within each air quality region are key requirements under all federal air quality legislation since 1970. In general, these requirements are delegated to the respective states, since they, in general, have a better understanding of the sources of contaminants and are responsible under separate legislation for providing monitoring, inspection, and enforcement.

The federal government has in effect *new source performance standards* (NSPS) requirements for new sources in specified industries which are required to meet nationwide emission standards. The focus of the NSPS requirements are criteria pollutant emissions from the largest stationary source categories in the country. These include fossil fuel-fired electric utility generating plants, portland cement plants, nitric and sulfuric acid plants, petroleum refineries, asphalt concrete plants, secondary metal smelters, iron and steel plants, fertilizer plants, etc.

One of the dominant areas of federal authority is the setting of "tailpipe" emission standards for motor vehicle emissions, as well as overseeing fuels and additives for those sources. This is in recognition of the fact that mobile sources emit the criteria pollutants NO_x, carbon monoxide, and ozone precursor hydrocarbons.

Prevention of Significant Deterioration

Federal regulations also require that the air quality does not deteriorate further in those areas where the air is already "cleaner" than the NAAQS. Under the *Prevention of Significant Deterioration* (PSD) regulations, all the nation's air quality control regions with a NAAQS were divided into three classes of ambient air quality. Those regions which were experiencing air quality significantly below the ambient air quality standard were designated Class I. Therefore, only very small incremental increases of air contaminants are permitted in those areas. Class II areas are those where the air pollution is in excess of national standards, and different levels of incremental addition to existing air contaminant levels are allowed. No Class III areas have been designated to date.

The PSD program initiated the concept of *new source review* (NSR). This emissions concept was one in which limited degrees of incremental additional air pollutants were allowed in the air quality regions. Stationary sources were the focus. In order for a major source to be built, the new source in a PSD-regulated area was not allowed to increase the existing total emissions in that area.

The concept of *offsetting* emissions was established. In this approach, other emissions were required to be reduced at a ratio equal to *or greater than* the anticipated new emissions prior to the construction of a new source of air contaminants. The allowable amounts were established using modeling as a planning tool.

The concept of *lowest achievable emission rate* (LAER) was established for stationary sources. LAER is the degree of emissions control considered the most stringent for a source in a nonattainment area, and applies to new or modified major sources. The definition of major source or major modification depended upon the contaminant. LAER was based upon the most stringent rate contained in any state implementation plan, or that level of control achieved in practice by similar sources.

For those sources in areas attaining the ambient air quality standard another level of control technology is established. This is best available control technology (BACT). BACT is determined on a case-by-case basis for new sources in PSD areas and takes into account energy and economic, as well as environmental impacts. While more flexible than LAER, BACT always has to be sufficient to meet new source performance standards.

Emergency Episodes

In addition to those regulations at the federal level dealing with ambient air quality, the federal government was given the authority to deal with air quality episodes. These included stringent limitations on operations, emissions, fuel use, etc., in the event of federal air quality emergency levels (Chapter 3) being exceeded. Area wide shut-downs of industrial and commercial operations are also allowed under federal emergencies.

Hazardous Air Pollutants

In addition to concerns for criteria pollutants, the federal government established national emission standards for hazardous air pollutants (NESHAPs). The NESHAP regulations established nationwide standards for existing, modified, or new sources that emitted any one of the original seven listed hazardous air pollutants (HAPs). These air contaminant emissions were limited to certain specific industries which emitted those NESHAP pollutants. The NESHAP concept is expanded on in the Clean Air Act Amendments (CAAA) of 1990.

Global Concerns

The EPA was given the authority to regulate stratospheric ozone-depleting chemicals in earlier versions of the Clean Air Act. This was the first attempt at dealing with emissions of a global nature. The focus was protection of the stratospheric ozone layer based upon early research indicating a link between potential ozone destruction and the emissions of certain chlorofluorocarbons (CFCs). The power to regulate these emissions was significantly expanded upon in the latest Clean Air Act Amendments.

Other Federal Acts

The federal government has the authority to regulate air emissions under a variety of different federal laws in addition to the CAA. These include other laws dealing with operations or activities which may generate air emissions. These are typically fugitive emissions such as volatile organic compounds or hazardous air pollutants as well as criteria contaminants, such as particulates and NO_x.

The Toxic Substances Control Act (TSCA) was the first regulation to deal with air contaminant emissions of a hazardous nature by its regulation of emissions of polychlorinated biphenyls (PCBs). Incinerators discharging air contaminants while burning PCB and PCB-containing waste materials were required to meet a strict level of control efficiency. The *destruction and removal efficiency* (DRE) for PCBs in the exhaust gases had to be equal to or greater than 99.9999%. This was based upon the total PCB mass input to the incinerator.

The Resource Conservation and Recovery Act (RCRA) deals with ongoing waste management facilities which have an air emissions component, or which deal with waste fuels. The emissions from facilities handling waste materials may be fugitive as well as direct or combustion-oriented for their air component.

RCRA established specific standards for incinerators disposing of hazardous waste by requiring, under the provisions of a "trial burn" (for the operating permit), that the DRE be equal to or greater than 99.99% for the principal organic hazardous compounds (POHCs) identified in the waste materials. In addition, a limitation on the emissions of hydrochloric acid, particulates, and CO were established for the operating permits for those RCRA incinerators.

Whenever any "corrective action" is required at a RCRA facility, the owner or operator of that facility must deal with all air emissions occurring as a result of that corrective action. Such corrective action may be decontamination of soil and/or ground water during a cleanup activity at the ongoing waste facility.

Recent RCRA regulations deal with air emissions from *hazardous waste* burned in *boilers and industrial furnaces* (BIFs). For these rules, the facility must obtain a permit under RCRA to burn such waste fuels. With respect to organic emissions, the BIFs must meet the DRE standards of 99.99% for all listed waste materials and their fuels, and a 99.9999% DRE for those wastes which contain dioxin. These boilers and industrial furnaces are also subject to emission limits for certain heavy metals, HCl, and chlorine gas, particulates (0.08 grains per standard dry cubic foot at 7% oxygen), and carbon monoxide. While the majority of the focus is on the air quality side, these are regulations under RCRA.

Hazardous waste site cleanups (where no current operator or owner exists) are regulated under the Comprehensive Emergency Response, Compensation and Liability Act (CERCLA) commonly known as Superfund. The amendments of 1986, termed the Superfund Amendments and Reauthorization Act (SARA) further clarified the requirements for such cleanup activity.

Under CERCLA/SARA, all existing federal regulations for either NSPS or NESHAPS must be met for any cleanup activity. In addition, concerns for fugitive emissions, monitoring of the air at the perimeter of the facility during cleanup activities, emissions testing of sources during remediation, as well as public input, must be provided for during the implementation of site restoration.

For a CERCLA/SARA cleanup action, federal authority preempts all local and state regulations for air quality management; however, remedial actions must take into account all local *applicable, relevant and appropriate regulations* (ARARs) during cleanup activities. No permit is required for cleanup of a federal "superfund" site, since the remediation is carried out under the authority of the federal government. The rationale is that, due to the immediate health risk of hazardous waste, specific regulations requiring long periods of time (such as permitting) are preempted by federal authority.

THE 1990 AMENDMENTS TO THE CLEAN AIR ACT

The signing of the Clean Air Act Amendments (CAAA) on November 11, 1990, introduced sweeping changes in the federal approach to air quality management. The Clean Air Act Amendments (CAAA) redirected the entire scope of federal regulations with respect to criteria pollutants, hazardous air pollutants, and global issues.

There were 11 major titles to the CAAA, and these titles and the significant provisions and focus of each are seen in Table 9.1. Some of the new provisions deal with ambient air quality standards, changes in mobile source regulations, hazardous air pollutants, acid deposition, federal permits, stratospheric ozone protection, enforcement, and a number of miscellaneous and research provisions. In addition, it outlined a 20 year time frame for regulations to be adopted in order to implement the specifics of the CAAA. Each of these are reviewed in the following sections. Key among these is the emphasis on attainment of ambient air quality standards and protection of public health and welfare.

For each of the specific titles, a synopsis of the major provisions is provided below. These will follow the specific titles of the amendments and the major focus of each. Table 9.2 summarizes the *time schedule* for the major provisions of the Amendments.

Title I – Attainment and Maintenance of the NAAQS

A recognition that the ambient air quality standards were not being met in a timely fashion led to the major provisions of Title I. In particular, these concerned ozone air quality, ozone formation, oxides of nitrogen and hydrocarbon emissions, and the relationship of fuels and combustion to CO emissions.

Title I classifies and assigns attainment dates for ozone, carbon monoxide, and PM10 for different areas of the country. For each classification, specific measures for implementation and sanctions are outlined.

Using the statistical metropolitan area (SMA) Census Bureau tabulations, as well as the air quality control regions, the EPA organized five classifications for those areas in the U.S. which were not attaining the ozone standard and two classifications of nonattainment for both carbon monoxide and for PM10.

The designations of classes range from "marginal" through "extreme" nonattainment areas. These classifications are based upon the *design value*, which is the fourth highest contaminant concentration (usually ozone hourly average) monitored anywhere within any one of those regions in a three consecutive year period. Based upon the design value, areas were classified, and specific requirements were listed in Title I for each of those areas.

Table 9.3 illustrates the three contaminants, the classifications, the design

Table 9.1. Significant Provisions of the Clean Air Act Amendment

I	Attainment and Maintenance of the NAAQSs
	• Classification and Attainment Dates
	• SIP Revision
	• NO_x Requirements
	• Multistate Areas and Sanctions
II	Mobile Source Provisions
	• Vehicle Emission Standards
	• Emissions Control and Compliance
	• Fuel Requirements
	• Non-Road Engines
	• Reformulated and Oxygenated Gasoline
	• Clean Fuels
III	Hazardous Air Pollutants
	• Pollutant Lists and Source Categories
	• Emission Standards and Compliance Schedules
	• State Programs
	• Shoreline Deposition
	• Special Studies
	• Prevention of Accidental Releases
	• Risk Assessment and Management Commission
	• Solid Waste Combustion
IV	Acid Deposition Program
	• SO_2 Provisions
	• NO_x Provisions
	• Emissions Documentation
	• Clean Coal Technologies
V	Permits
	• EPA/State Program Interface
	• Program Requirements
	• Permitting Process
	• Special Provisions
VI	Stratospheric Ozone Protection
	• Chemical Lists and Ozone Depletion Potentials
	• Reporting
	• Reduction and Schedules
	• Use, Recycling and Disposal of Chemicals
	• Labeling of Products made with Ozone Depleting Chemicals
VII	Enforcement Provisions
	• Administrative, Civil and Criminal Provisions
	• Judicial Review
	• Citizen Suits
VIII	Miscellaneous
	• Outer Continental Shelf Emissions Sources
	• Contaminated Oil-Ship Fuel
	• Visibility and Source/Receptor Concepts
	• Interstate Commission
	• Fuel Cell Research
IX	Clean Air Research
	• NAPAP/Acid Rain Research Continuation
	• Liquified Gaseous Fuel Spill Dispersion Test Facility
	• Clean Alternative Fuels
X	Disadvantaged Business Concerns
XI	Clean Air Employment Transition Assistance

Table 9.2.　Time Line for Major Clean Air Actions by States and EPA

1993 •	Affected facilities in ozone nonattainment areas file annual VOC and NO_x emission statements.
•	Reports to Congress on VOC emissions from consumer and commercial products.
•	EPA issues emission standards for nonroad engines and vehicles.
•	Diesel fuel must meet 0.05% by weight sulfur standard.
•	States submit proposed permit programs to EPA.
1994 •	EPA issues emission standards for waste-fueled boilers and industrial furnaces.
•	EPA must approve or deny state permit programs.
1995 •	Moderate CO nonattainment areas must achieve standard.
•	Some major source categories emitting hazardous air pollutants start to meet MACT standard.
•	States must have approved permit program or EPA will administer.
•	Sources start to file permit applications in some states.
1996 •	Moderate ozone nonattainment areas must achieve standard.
•	Ozone nonattainment areas from moderate on up must reduce VOC emissions by 15%.
•	States start processing and issuing permits for sources.
•	EPA submits report to Congress concerning the assessment of public health risks associated with residual air emissions.
•	Employers required to comply with transportation control measures adopted by states in serious, severe and extreme ozone nonattainment areas.
1998 •	New heavy duty trucks meet 4.0 gbh NO_x emission standard.
•	Centrally-fueled fleets of heavy duty trucks must purchase some clean fuel vehicles.
1999 •	Serious ozone nonattainment areas must achieve standard.
2000 •	EPA completes the issuance of MACT standards for all major sources emitting hazardous pollutants.
•	Serious CO nonattainment areas must achieve standard.
•	EPA may start to issue risk-based emission standards for major sources.
2005 •	Severe ozone nonattainment areas must achieve standard.
2010 •	Extreme ozone nonattainment areas must achieve standard.

level (in parts per million), and the attainment dates assigned to each class by which time the ambient air quality standard must be reached. For each area, there are periods from 3 to 20 years from the date of enactment by which the implementation of specific air quality management strategies and regulations must be implemented, and by which date the ambient air quality standard must be obtained.

In Tables 9.4 and 9.5 are the ozone and carbon monoxide classifications

Table 9.3.   Title I Nonattainment Classifications and Levels

	Class	Level — ppm	Attainment Date
Ozone	Marginal	.121 to .137	1993
	Moderate	.138 to .159	1996
	Serious	.160 to .179	1999
	Severe 1	.180 to .190	2005
	Severe 2	.191 to .279	2007
	Extreme	.280 and above	2010
Carbon Monoxide	Moderate	9.1 to 16.4	1995
	Serious	16.5 and up	2000
For ozone and CO: EPA may grant two 1-year extensions of attainment date			
PM-10	Moderate	N/A	12/31/94 6 years for future areas
	Serious	N/A	12/31/01 10 years for future areas
Possible extension of attainment date up to 5 years for serious areas			

by city area which are violating the NAAQS from the "moderate" through "extreme" nonattainment classifications. Marginal areas for ozone included 39 metropolitan areas (not listed) which were to have achieved attainment by November 1993. As seen earlier, these marginal areas are only slightly above the ambient air quality standard. Moderate through the extreme classification areas are therefore far enough above the ambient air quality standard that significant air quality management strategies have to be implemented.

Ozone Nonattainment Requirements

Table 9.6 outlines the major mandatory strategies for each of the ozone nonattainment areas. For the moderate, serious, and severe areas of nonattainment for ozone, there will be an increasing gradation of severity of measures required in order to meet the ambient air quality standard. The Los Angeles basin, as the single "extreme" area, is in a class by itself. For each of these areas, a series of mandatory changes to the State Implementation Plans are required, depending on the degree of nonattainment. In each case, in moving from the moderate to the extreme cases, additional measures are required.

Table 9.4.   Ozone Nonattainment Areas

Moderate Areas	Serious Areas
Atlantic City, NJ	Atlanta, GA
Bowling Green, NY	Bakersfield, CA
Charleston, WV	Baton Rouge, LA
Charlotte, NC-SC	Beaumont, TX
Cincinnati, OH-KY-IN	Boston, MA-NH[a]
Cleveland, OH[a]	El Paso, TX[a]
Dallas, TX	Fresno, CA[a]
Dayton-Springfield, OH	Hartford, CT[a]
Detroit, MI	Huntington, WV-KY-OH
Grand Rapids, MI	Parkersburg, WV-OH
Greensboro, NC[a]	Portsmouth, NH-ME
Jefferson Co., NY	Providence, RI
Kewaunee Co., WI	Sacramento, CA[a]
Knox Co., ME	Sheboygan, WI
Louisville, KY-IN	Springfield, MA
Memphis, TN-AR-MS[a]	Washington, DC-MD-VA[a]
Miami, FL	
Modesto, CA[a]	**Severe Areas**
Nashville, TN	Baltimore, MD[a]
Pittsburgh, PA	Chicago, IL-IN-WI
Portland, ME	Houston, TX
Raleigh-Durham, NC[a]	Milwaukee, WI
Reading, PA	Muskegon, MI
Richmond, VA	New York, NY-NJ-CT[a]
Salt Lake City, UT	Philadelphia, PA-NJ-DE-MD[a]
San Francisco-Oakland-San Jose, CA[a]	San Diego, CA[a]
Santa Barbara, CA	
St. Louis, MO-IL	**Extreme Area**
Smyth Co., VA	Los Angeles, CA[b]
Toledo, OH	
Visalia, CA	
Worcester, MA	

[a]Moderate CO nonattainment.
[b]Serious CO nonattainment.

Marginal Areas

For the marginal areas, accurate emissions inventories must be completed, new source review requirements will be required on major NO_x sources, and automotive inspection and maintenance (I/M) of mobile source control systems will be required. Major stationary sources of VOCs and NO_x must submit inventories of emissions beginning in 1993 and every year thereafter.

An offset ratio for new stationary source emissions of at least 1.1 is

Table 9.5.   CO Nonattainment Areas Only

Moderate Areas	
Albuquerque, NM	Missoula Co., MT
Anchorage, AK	Phoenix, AZ
Chico, CA	Portland-Vancouver, OR-WA
Colorado Springs, CO	Provo-Orem, UT
Denver-Boulder, CO	Reno, NV
Duluth, MN-WI	Seattle-Tacoma, WA
Fairbanks Ed., AK	Spokane, WA
Fort Collins-Loveland, CO	Stockton, CA
Josephine Co., OR	Syracuse, NY
Klamath Co., OR	**Serious Areas**
Las Vegas, NV	Steubenville-Weirton, OH-WV
Medford, OR	Winnebago Co., WI
Minneapolis-St. Paul, MN-WI	(Oshkosh)

required (110% reduction). For each increase in *VOC emissions*, there must be an equivalent 115% decrease under new source review regulations under the CAAA.

Moderate Areas

In addition to the requirements for marginal areas, the moderate areas of nonattainment must revise the state implementation plans to include, for *mobile sources,* requirements for additional air pollution control, inspection and maintenance programs, and Stage II vapor recovery systems during refueling (vacuum-assisted vapor recovery at the nozzle). Contingency measures are to be included in the event that these requirements fail to attain the ambient air quality within the six-year time frame. *Reasonable further progress* requirements for the moderate ozone areas will include up to a 15% total *VOC reduction* in the entire area affected.

For stationary sources of VOCs and NO_x, the emission *threshold* for inclusion in SIP revisions is defined as 100 tons per year for new and existing plants. Forty tons per year is the emission threshold for increases during a *modification* of an existing facility.

For *control systems* on the major sources, the reasonably available control technology (RACT) is required for the existing major sources of VOCs and NO_x. For new major sources or modifications of an existing major source, the lowest achievable emission rate (LAER) level of control will be

Table 9.6. Mandatory Strategies for Ozone Nonattainment Areas

Strategy	Moderate	Serious	Severe	Extreme
Mobile Sources:				
	Inspection and Maintenance	Clean Fuel Fleet Programs	Total VMT Reductions- TCMs	Enhanced Lower Measures
	Refueling Vapor Recovery or On-Board VR	Transportation Control and Congestion Measures	100 Employees AVOR of 1.25	Traffic Controls
	Contingency Measures	Fleet Refueling Provisions		Contingency Measures
		Enhance Inspection & Maintenance: Emissions Diagnostics Annual Testing Registration Denials		New Technologies
Stationary Sources:				
Reasonable Further Progress	15% VOC reduction	NO_x and VOC reductions	–	–
VOC & NO_x Thresholds, TPY				
New & Existing Major Facility	100	50	25	10
Major Modification	40	25	25	any

Control System Levels				
Existing Majors	RACT	RACT	RACT	RACT
New or Major Modifications	LAER	RACT or BACT[a] / LAER	LAER	LAER
Large Boilers	—	—	—	[a]Clean Fuels or New NO_x Controls
NSR VOC Offset Ratios				
Internal	1.0	1.3	1.3	1.3
External	1.15	1.2	1.3	1.5

Each category automatically includes all strategies for lower levels.

[a]BACT is for modificaitons for <100 TPY sources when emission increases are not offset (Best Available Control Technology)

AVOR = Average Vehicle Occupancy Rate
RACT = Reasonable Available Control Technology
LAER = Lowest Achievable Emission Rate
TPY   = Tons Per Year
VMT   = Vehicle Miles Traveled
VOCs = Volatile Organic Compounds
VR   = Vapor Recovery during refueling operations

required. For new source reviews, *internal* "offset" ratios of at least 1.0 are required. *External* offsets (those found outside the plant boundaries) are 115% of the VOC increases at the new facility.

Serious Areas

For the serious ozone nonattainment areas, mobile sources receive much greater scrutiny. These include requirements for clean fuel programs for fleet vehicle owners, transportation and congestion management plans or control measures for all mobile sources, vapor recovery requirements for fleet owners during vehicle refueling, and "significantly enhanced" inspection and maintenance procedures. The latter require annual auto emissions testing, repair and maintenance, provisions for denial of registration for vehicles failing the test, and decentralized testing and certification of on-board emissions control diagnostics systems and computerized emission analyzers.

Reasonable further progress requirements include NO_x and VOC reductions. The threshold levels (existing emissions) for VOC and NO_x stationary sources are 50 tons per year for new or existing facilities, and 25 tons per year for an existing source.

Control levels required for those major new or modified stationary sources are at the *lowest achievable emission rate*, whereas for existing sources either RACT or BACT (best available control technology) may be utilized. BACT applies in those cases where a modified facility has no offsets for its emission increases and when the total emission is less than 100 tons per year of VOCs. The VOC offset ratios under new source review will be 120% for those sources outside plant boundaries, whereas they will be 130% (or 1.3 to 1) for internal sources of VOCs.

Severe Areas

In addition to all of the preceding requirements on mobile and stationary sources, those areas classified as *severe nonattainment* will see requirements for mobile sources to include total vehicle miles traveled (VMT) reductions by the imposition of transportation control measures. Employers of more than 100 persons will be required to implement incentives to increase the average vehicle occupancy rate (AVOR) to at least 25% above existing.

The VOC and/or NO_x new source review threshold for all facilities becomes 25 tons per year, and the control level becomes LAER for major new or modified sources. The VOC offset ratios under new source review are 1.3 to 1 for both internal and external reductions of emissions.

Extreme Area

Los Angeles, the only *extreme ozone nonattainment* area, will be required to implement further controls on all sources. For mobile sources, in addition to all of the preceding requirements, there will be demands for new technologies, additional traffic control measures during heavy traffic hours, and enhancement of all of the lower level requirements. Further contingency measures are required in the event that those fail. These are to be included in the state implementation plan as mandatory requirements.

The VOC and NO_x threshold for the definition of a major source is 10 tons per year for a new or existing facility or any increase during modification of an existing source.

Again, the federal control level is LAER for all major new or modified sources with reasonably available control technology (RACT) for all existing major sources as a *retrofit* requirement. This is the minimum, quite apart from the additional controls imposed by the state or local authorities.

Additional controls are imposed on all combustion systems for boilers; "clean fuel" combustion or new and additional NO_x emission controls are specifically required. The VOC offset ratios for new source review are 1.5 for external sources or 1.3 for internal sources controlled at a given facility.

Additional Ozone Strategies

Title I provides for a number of additional strategies to be implemented in all ozone nonattainment areas. The first of these is the requirement that all of the plan provisions for stationary VOC sources also apply to major stationary sources of NO_x, unless the EPA determines that additional NO_x controls would not create a net benefit, or for areas which are not a part of an *ozone transport region*.

Milestones

Milestones are required under Title I for serious, severe, and extreme nonattainment areas to demonstrate that there have been applicable emission reductions met within stated time periods. In addition, for the areas which fail to demonstrate compliance with those milestones, there are additional requirements that the EPA adopt, or force the adoption, of an "economic incentive program" for those areas which fail to meet those compliance demonstration requirements. In addition, those areas may be reclassified up to the next higher classification and/or be required to implement control measures adequate to meet the next milestone in emission reductions.

Interstate Transport

Recognizing that certain geographic areas are specific sources of contaminants leading to ozone nonattainment and other areas may be merely the receptor, Title I allows for the inclusion of a multi-state area to be set up, in which the included states must coordinate SIP revisions. In addition, Title I requires the use of photochemical grid modeling, similar to the EKMA model, to provide the analytical tools to evaluate SIP revisions and new requirements.

A specific ozone transport region and a commission for one region has been specifically required in the law. This includes the area from Washington D.C. northward along the eastern Atlantic seaboard to Maine. These 11 states and the District of Columbia, under a transport commission, must assess the degree of interstate transport and recommend measures to the EPA necessary to ensure that the relevant SIPs meet the plan requirements.

Additional requirements for these states include enhanced vehicle inspection in metropolitan areas of greater than 100,000 population, RACT levels of control in all sources of VOCs, and Stage II vapor recovery controls during vehicle refueling. Stationary sources emitting at least 50 tons per year of VOCs are considered major by this provision of Title I and are subject to all plan requirements applicable to at least a moderate nonattainment area. EPA retains the right to oversee all activities of the interstate transport region.

Control Guidelines

The EPA is required to establish *Control Technique Guidelines* (CTGs) for 11 additional categories of stationary source VOC emissions. These must give priority to those categories that make the most significant contribution to ozone nonattainment. These also must include hazardous waste treatment storage and disposal facilities (TSDFs). Specifically named as requiring CTGs are aerospace coatings and solvents, and emissions of

VOCs and PM10 from paintings, coatings, and solvents used in ship building and repair. *Best available control measures (BACM) are required in these CTGs.*

Consumer Products

Consumer or commercial products are required to be addressed by the EPA. By November 1993, the EPA must list categories of consumer or commercial products that account for at least 80% of the VOC emissions from such products in ozone nonattainment areas, and must require best available controls.

The EPA may control or prohibit by regulation the manufacture, sale, or introduction to commerce of any product that is a source of VOC emissions. In addition, the EPA may impose fees, charges, or collect funds associated with the regulations of these products in addition to requiring labeling, self-monitoring and reporting, prohibitions and limitations on these materials.

Marine and Harbor Emissions

VOC emissions and any other pollutants from the loading and unloading of marine tank vessels in harbors in ozone nonattainment areas are the focus of additional standards which will require reasonably available control technology (RACT) for such emissions.

Additional Fees

In addition to all of the above, stationary sources in those areas classified as *severe or extreme* which fail to achieve attainment by their respective deadlines, will be required to *pay an annual fee* to the state beginning the year after the attainment date.

The baseline amount begins at $5,000 per ton of VOCs emitted during a calendar year. If the state fee provisions are not adequate in the SIP, the EPA may collect the unpaid fees. These fee provisions do not apply if the population is less than 200,000 and the nonattainment area is a receptor of ozone transport.

Additional sanctions against ozone nonattainment areas include prohibitions on highway funding, withholding of air pollution planning or control grants, and a requirement that all emissions be offset by at least 2 to 1 for any increases under new source review.

Carbon Monoxide Nonattainment Provisions

The mandatory Title I provisions for CO nonattainment areas closely parallel many of the provisions for ozone nonattainment. The focus is combustion sources of emissions, primarily from vehicular sources. These provisions include restrictions on vehicle miles traveled (VMT), submissions of accurate current *inventories* of CO emissions, inspection and maintenance programs of mobile source control systems, clean fuel fleet programs, and transportation control measures as required in the severe ozone nonattainment areas, except that the program applies to CO.

Enhancements of each of the mandatory mobile source control requirements are required for the serious nonattainment areas. For both the *moderate and serious* areas of nonattainment, oxygenated fuels (oxygen content not less than 2.7%) are required in all motor vehicle fuels during those times of the year that are considered "high CO conditions."

If it is determined that stationary sources in serious areas contribute significantly to CO levels, a major stationary source is defined as one which has the potential to emit 50 tons per year. It will be subject to major source requirements for CO similar to those for ozone.

In addition, an economic incentive and transportation control program will be required for each state if they fail to meet their milestone demonstration on time. The economic incentive program, when combined with other revisions in the SIP, must reduce total CO emissions in the area by 5% each year until attainment. Within the three serious areas, the oxygenated fuel requirement is a minimum of 3.1% by weight, as required under this provision.

PM10 Nonattainment Areas

Initially, all areas considered nonattainment for PM10 were classified as moderate. Upgrades to serious areas of nonattainment are provided for in Title I.

For those stationary area sources in the moderate areas, states must implement a construction and operating permit program (if it does not already exist) for new and modified major stationary PM10 sources. Reasonably available control measures (RACM) for those PM10 sources are to be implemented by December 1993, or four years after designation as a moderate area of nonattainment.

In the serious PM10 nonattainment areas, a major source is defined as one that emits at least 70 tons per year of PM10. For those sources of PM10, best available control measures (BACM) are to be implemented not later than four years after the area is classified as serious. All of the plan requirements for moderate areas apply and, in addition, the states must provide a demonstration of attainment by the most expeditious alternative date practicable by use of air quality modeling.

If a serious area does not reach attainment by the deadline, SIP revision must be submitted which allows for an additional annual 5% emission reduction of PM10 or PM10 precursor emissions in the area, as reported in the most recent inventory. In addition, PM10 control measures for stationary sources also apply to major stationary sources of PM10 precursors.

The EPA is required to issue technical guidance on what constitutes RACM and BACM for urban fugitive dust sources, residential wood combustion, and prescribed agriculture and forestry-clearing burning operations.

Title II — Mobile Source Provisions

As seen in Chapter 4 (sources), transportation emissions by light-duty and heavy-duty vehicles and trucks are among the most significant contributors to air pollutants. In Title II, the CAAA details requirements for all aspects of new regulation over mobile source emissions. These relate to both emission standards as well as the fuels utilized in those programs and vehicles. The first of these takes the form of tailpipe emission standards, whereas the latter take the form of requirements for fuel and fuel compositions.

Light-Duty Vehicle Standards

Table 9.7 details the emission standards for light-duty vehicles for the NO_x, CO, nonmethane hydrocarbon and particulate matter emissions for both gasoline and diesel-powered light-duty vehicles. Increasing percentages of light-duty vehicles must meet the standards as a function of model year.

A significant new addition to the EPA requirement is that these standards be met for the initial 50,000 miles of travel and, additionally, a slightly higher level of emissions at the 100,000-mile mark. Heavier weight trucks

Table 9.7.   Title II Vehicle Emission Standards, Light-Duty Vehicles and Trucks Up to 6,000 Pounds Gross Weight

Pollutant	Weight (lb)	Emission Standards (gm/miles)	
		5 yr/50,000 mi	10 yr/100,000 mi
Gasoline:			
NMHC	0–3,750	0.25	0.31
	3,751–5,750	0.32	0.41
CO	0–3,750	3.4	4.2
	3,751–5,750	4.4	5.5
NO_x	0–3,750	0.4	0.6
	3,751–5,750	0.7	0.97
Diesel:			
NO_x	0–3,750[a]	1.0	1.25
	3,751–5,750	–	0.97
PM	LDVs & LDTs	0.8	0.10

[a]Prior to 2004 A.D.

have different requirements. Taken together, these new tailpipe standards are estimated to reduce hydrocarbons about 40%, and NO_x by about 60% from transportation sources.

Nonroad fuels and engines will be the focus of additional standards to achieve the greatest degree of emission reductions achievable. By 1995, the EPA must promulgate separate standards for new locomotives and the engines powering them. Again, the greatest degree of emissions reduction achievable, using available technology, is required.

California is expressly authorized to regulate nonroad engines and vehicles with standards that are at least as protective as federal standards, whereas other states are not allowed that waiver of federal preemption. However, any state with an approved SIP may adopt the California approach as long as they are equivalent.

Urban buses under Title II must achieve a 50% reduction in particulate matter. Additional low-polluting fuel requirements are phased in over a period of five years for urban transit buses.

The EPA CO emissions standard during cold-start conditions (20°F) is 10 grams per mile for 1994 and afterward. These are phased in by sales percentage for vehicle manufacturers.

In addition to tailpipe emissions, evaporative emissions controls are required from gasoline vehicles operating in summer high ozone conditions during operation and following two or more days of no usage. These regulations must include the greatest degree of emission reduction achievable considering volatility, cost, energy, and safety factors.

Tailpipe Toxics

With respect to mobile source air toxics control, the EPA is required to establish a minimum emission standard for benzene and formaldehyde from all mobile sources in 1994. These standards must reflect the greatest degree of emission reduction achievable using available technology, considering cost and availability of technology, noise, energy, safety, and lead time.

Emissions Control

Two key issues regarding nontailpipe standards revolve around emissions during refueling and computerized diagnostic systems. These latter evaluate the ability of existing air pollution control systems on vehicles to function properly. The first requirement is that new light-duty vehicles will require on-board vapor recovery systems with a minimum capture efficiency of 95%.

By 1994, 40% of all manufacturers' sales volume must meet the standard, with the sales volume rising to 80% by 1995, and 100% of all vehicles having on-board vapor recovery systems by 1995. For serious, severe, or extreme ozone nonattainment areas, the EPA may require these *in addition to* current Stage II vapor-recovery requirements during vehicle refueling.

In addition to standards upon sale, vehicles are required to have their own on-board *computerized diagnostic systems* which will evaluate and alert drivers as to whether these on-board control systems are still being effective. Beginning in 1994, all light-duty vehicles and trucks must have these on-board emission control diagnostics.

States must include in their SIPs that the inspection and maintenance programs will include evaluation and testing of such on-board systems. This will assure that control systems will remain effective on each individual vehicle, rather than a general average being computed for all vehicles in the system.

Automobile warranties must cover these diagnostic systems as well as catalytic convertors and emission control systems for at least eight years or 80,000 miles, beginning in 1995. *Penalties* at the federal level for tampering with emission-control devices vary from $2500 to $25,000 for tampering with, defeating, or rendering inoperative such control systems.

Fuel Requirements

Specific fuel compositions will have direct impacts on air emissions from combustion of such fuels. Therefore, Title II specifically looks at fuels and fuel composition to help attain ambient air quality standards. Specific items which come under the heading of fuel requirements relate to prohibitions for mis-fueling of leaded gasoline in vehicles intended only for unleaded gas. Another requirement is that the Reid Vapor Pressure (RVP) not exceed 9.0 pounds per square inch (psi) unless the fuel blend is for gasoline with 10% ethanol, in which case the RVP requirement is 10.0 psi.

With respect to diesel fuels, the sulfur content may not exceed 0.05% (500 ppm) after October 1993. Beginning January 1, 1996, all motor vehicle gasolines containing lead or lead-based additives will be totally prohibited nationwide. The manufacture, sale, or introduction to commerce of any motor vehicle which requires leaded gasoline was prohibited in 1992.

Additional fuel requirements are in place for the nine worst ozone nonattainment areas with a 1980 population greater than 250,000. These requirements include reformulated and oxygenated fuels. A reformulated fuel must meet certain general requirements for NO_x, oxygen, benzene, and heavy metal content and must also achieve reductions in ozone-forming VOCs and toxic air pollutants.

Table 9.8 details the minimum composition requirements for reformulated fuels. These include a minimum 2.0% by weight oxygen content, a benzene level of 1.0% or less by volume, and a prohibition on any heavy metals, including lead or manganese. The aggregate limit of aromatic hydrocarbons is 25% by volume, and a use of additives is required as needed to meet VOC and toxic emission standards. In addition, the emission of ground level ozone-forming compounds (ozone VOCs) and toxic air-pollutants must be reduced by 25% and 20%, respectively, from the aggregated baseline levels.

The CO nonattainment areas must use oxygenated fuels that have a minimum content of 2.7% oxygen by weight. These are required to be sold during the high CO portion of the year (typically wintertime conditions). Since the combustion of these reformulated gasolines can cause increased emissions of NO_x, the new law also prohibits vehicles which are using reformulated gasoline from increasing NO_x emissions beyond the levels associated with vehicles burning conventional fuels.

In addition to varying the percentages of standard gasolines, Title II lays out a requirement for fleet vehicles (10 or more) to use "clean fuels" in the serious, severe, or extreme ozone nonattainment areas. These clean fuels

Table 9.8.   Composition of Baseline and Title II Reformulated Gasolines

	Baseline Gasoline	Reformulated Fuel
Constituent, %		
Oxygen	0	2.0 (min)[b]
Benzene	1.53	1.0 (max)
Lead	[a]	0.0
Aromatics	32	25 (max)
Sulfur, ppm	339	—
NO_x	—	No increase
Reductions (2000 A.D.), %:		
Ozone VOCs	—	25
Toxics[c]	—	20 (min)

[a]Not applicable.
[b]2.7% (min) in CO nonattainment areas for Oxygenated Fuels.
[c]Total emissions of: benzene, formaldehyde, acetaldehyde, 1,3 butadiene, and polycyclic organic matter (POM).

include methyl alcohol, ethanol, blends of other alcohols of 85% with gasoline, reformulated gasolines or diesels, natural gas, liquefied petroleum gas, hydrogen, or other power sources. Beginning in 1998, 30% of all new fleet vehicle purchases must be utilizing "clean fuels."

As a pilot program for clean-fueled vehicles, the EPA is establishing a program in California which requires at least 150,000 vehicles to be produced, sold, and distributed there by 1996, 1997, and 1998, with 300,000 vehicles in the state by 1999. The experience of California will then become a laboratory for utilizing clean fuels and ultra-low emission vehicles for fleet systems. Other states may "opt in" to the California pilot program, provided they are in a serious, severe, or extreme ozone area, and revise their SIP.

The California low-emitting vehicle (LEV) standards are 0.075 grams per mile of nonmethane hydrocarbons, 3.4 grams per mile of CO, and 0.2 grams per mile of NO_x. These California LEV standards are for clean fuel emission vehicles sold in California. Utilizing the clean fuel vehicle approach, ultra-low emission vehicles (ULEVs) and zero emission vehicles (ZEVs) are other techniques for fleet operators to obtain credits for those fleet vehicles not meeting the clean fuel requirements.

Table 9.9.   Title III Hazardous Air Pollutant Classes

Metal compounds:

• Antimony	• Arsenic	• Beryllium	• Cadmium
• Chromium	• Cobalt	• Lead	• Manganese
• Mercury	• Nickel	• Selenium	

Organics[a]:

• Benzene	• Biphenyl	• Carbon	• Methanol
• Dioxins	• Formaldehyde	Disulfide	• Toluene
• Nitrobenzene	• Phenol	• Hexane	
• Xylenes	• Halogenated Organics	• PCBs	
• Pesticides (2,4 D:	(TCA, TCE, etc.)		
DDE; Parathion,			
etc.)			

Acids, Oxidizers, and Physical Agents:

• Asbestos	• Chlorine	• Fine Mineral	• HF
• HCl	• Phosgene	Fibers	
• Radionuclides (includes	• Phosphorus		
Radon)			

[a]Some common representative substances.

Title III — Hazardous Air Pollutant Program

A major expansion of the Clean Air Act Amendments is in dealing with hazardous air pollutants (HAPs). Among the major new provisions of Title III, are:

- the listings of 189 hazardous air pollutants with source categories
- new levels of control technology (MACT) for HAPs
- provisions for area sources calculations of residual health risks after implementation of controls, and
- provisions for accidental releases of hazardous air pollutants.

The definition of a major source of hazardous air pollutants is *"a stationary source or group of stationary sources under common control which emit or have the potential to emit a total of 10 tons or more per year of any single hazardous air pollutants or 25 tons or more per year of any combination of HAPs."* For a single contaminant such as perchlorethylene, this might amount to as little as six and a half gallons per day being lost through fugitive emissions.

Some of the substances listed by the EPA in Title III are seen in Table 9.9. Substances may be deleted from or added to this list by the EPA based on additional scientific evidence regarding adverse human health effects. Specifically excluded from this list are the six criteria pollutants, including elemental lead. *Lead compounds,* however, are included on the list of hazardous air pollutants.

HAP Sources

The source categories are those industry groups which emit HAP substances. The categories are to be officially promulgated over a period of ten years (until the year 2000). Industrial source categories include cooling towers, electroplating operations, synthetic organic chemical manufacturing operations, decreasing operations, commercial sterilization facilities, dry cleaners, etc. Included are area sources such as gasoline stations, dry cleaners, manufacturers of furniture, printing presses, small boilers, etc. Also to be included are pulp and paper mills, petroleum refineries, polymer and resin operations, surface coatings, magnetic tape manufacturing, secondary metal operations, waste treatment storage and disposal facilities, and coke ovens.

Standards

Emission standards for HAPs are classified in two categories: technology based and health based. Emission standards must achieve the maximum degree of emissions reduction deemed achievable by the EPA for new or existing sources in the applicable category, considering cost, health, environmental impact, and energy requirements.

Technology Based Standards

The *maximum achievable control technology* (MACT) standards may include process changes, material substitutions, collection, capture and treatment of emissions, control technology, design equipment, and work practice operational changes, etc.

MACT is defined according to the amount of emissions from the HAP source. For new major sources, MACT is at least as stringent as the emissions level achieved in practice at a best controlled similar source. For existing major sources, MACT may be less stringent than that for new sources, but must be at least as stringent as either:

1. the average emissions limitation achieved by the best performing 12% of similar sources, (excluding those that have recently achieved the LAER for that category), or
2. the average emission limitation achieved by the best performing five sources, if the specific category has less than 30 units.

From smaller area sources, the EPA may require either MACT or a less stringent, *generally available control technology* (GACT).

Each source category has a deadline for attainment of the technology based emission standards. However, the new law does create an incentive for facilities to achieve *early reductions*. In this case, if all listed toxics are reduced at a facility (by 90% or more for organics and 95% or more for particulates), they may receive a six-year extension in the deadline to comply with MACT standards. This is to provide an incentive for early HAP reductions by whatever means.

Health Based Standards

Title III requires investigation of the health-based emission standards as applied to major sources. Area sources may also be included in the industry-related source category. These health-based standards must provide an ample margin of safety to protect public health and to prevent adverse environmental effects after implementation of MACT for major sources.

These health-based standards must also include the calculations of potential "residual risk" from *carcinogenic* air pollutants if the applicable technology-based standards are not reduced to a lifetime excess cancer risk of less than 1 in a million (individual risk) for the "maximum exposed individual."

If any single source shows a residual health risk greater than 1×10^{-6} (one chance in a million), the EPA must promulgate additional standards.

Area Sources

The EPA is to conduct a research program for area sources which yield risk categories accounting for 90% or more of the aggregate HAP emissions subject to these regulations. Specifically, the EPA must prepare a *national urban air toxics strategy* to reduce cancer risks from these area sources by 75%. This is in conjunction with other federal laws such as TSCA, FIFRA, and RCRA in the overall strategy of HAP reductions for health risks from area sources.

SIP Revisions

Again, under EPA authority, states must revise their implementation plans to set up HAP standards that are no less stringent than the federal requirements. In addition, states must review their statutes for regulation of potential high-risk point sources or "toxic hot spots." Such programs will also evaluate those facilities which produce, process, handle, or store substances listed under the accidental release provisions of Title III in quantities greater than threshold amounts. These revised state implementation plans are subject to review by the EPA.

HAP Permits

In addition to permits for major new or modified sources within a nonattainment area, permits are also required for air toxics sources. Title III hazardous air pollutant permits must: specify emission limits based on control technologies, require monitoring of emissions, and require compliance with baseline health-based emission limits. All major HAP sources must have permits.

Shoreline Deposition

Under Title III, the EPA is also required to institute a program to assess atmospheric deposition of HAPs into the coastal waters of the United States and around the Great Lakes. Such research is to involve monitoring and investigation of sources and deposition rates and research to evaluate adverse human health or environmental effects on biota, fish, and wildlife from hazardous air pollutants.

Special Studies

A number of special studies are required by Title III. These include a study of the emissions of hazardous air pollutants from electric utility steam-generating units. Alternative control strategies and potential regulations necessary to control emissions of HAPs from those sources are also studied.

Reports on mercury emissions from electric utility steam-generating units, municipal waste combustion units, and other sources, including area sources, are to be provided.

Special studies for coke oven production technology, publicly-owned treatment works (POTWs), and oil and gas wells are to be completed in the initial five to six years on HAP emissions and viable control technologies. In addition, special assessments on hydrofluoric acid and hydrogen sulfide emissions are to be provided by the EPA.

The National Academy of Sciences is to prepare, under arrangements with the EPA, a study on risk assessment methodology and improvements in that methodology.

Prevention of Accidental Releases

Title III sets up an accident prevention program similar to those in other federal statutes such as EPCRA, the Emergency Planning and Community Rights to Know Act.

The provisions of Title III for accidental releases apply to the 100 *extremely hazardous substances* originally listed under EPCRA. The following substances are included on the initial list which trigger the provisions of this section of Title III:

- chlorine
- hydrogen sulfide
- anhydrous ammonia
- toluene diisocyanate (TDI)
- methyl chloride
- phosgene
- ethylene oxide

- bromine
- vinyl chloride
- hydrogen fluoride
- methylisocyanate
- anhydrous sulfur dioxide
- hydrogen cyanide
- sulfur trioxide

The EPA has established *threshold release quantities* of those HAPs eligible for regulation at affected facilities.

For those facilities handling more than the threshold quantities of the initial 100 chemicals, *risk management plans* must be prepared by those facilities for dealing with accidental releases. These must include mitigation of the potential adverse human health or environmental effects from the release of more than the threshold amounts of HAPs. These plans must include a hazardous assessment, a program for preventing accidental releases, and a response program in the event of an accidental release. The risk management plan must be in accordance with guidelines issued by the EPA for those stationary sources, and must be registered with the EPA.

Regulations for the prevention and detection of accidental releases from these stationary sources must include use, operation, repair, replacement, maintenance of equipment to monitor, detect, inspect, and control releases of HAPs. The regulations also include provisions for personnel training in proper equipment operation to prevent releases.

Under Section 304 of Title III, the Department of Labor, in coordination with the EPA, must promulgate a *chemical process safety standard* to protect *employees* (pursuant to OSHA) from hazards associated with accidental releases of highly hazardous chemicals. The standard includes a listing of highly hazardous chemicals including toxic, flammable, highly reactive, and explosive substances.

A *risk assessment and management commission* has been set up as a "sunset" commission which provides investigation and recommendations on policy, implications, and appropriate uses of risk assessment and risk management in the federal regulatory programs to prevent cancer and other chronic human health effects.

A separate and distinct five-member *chemical safety and hazard investigation board* was established to investigate and report on facts, conditions, circumstances, and causes of any accidental releases resulting in fatality, serious injury, or substantial property damage. It also establishes regulations requiring reporting of accidental releases. In addition, the chemical safety board is authorized to conduct research and studies on the potential for accidental releases of extremely hazardous substances. This board is permanent and makes ongoing recommendations to the EPA or to the Secretary of Labor regarding accidental releases.

Solid Waste Combustion

Due to the concerns for emissions of hazardous air pollutants expressed in the recent past, such as dioxins from combustion of solid waste, Title III establishes a requirement for New Source Performance Standards (NSPS) for solid waste incineration units. Both existing and new or proposed solid waste units come under the provisions of Title III.

Emission standards must reflect the maximum degree of emissions reduction, considering cost and non-air quality health and environmental impacts. In addition, Title III sets up specific requirements to monitor emissions and report findings, provide operator training, and have a permit as issued under Title IV.

Acid gas scrubbers are required to be reviewed as a control technology for small new or existing facilities before promulgation of new source performance standards. Provisions are also made for handling ash from solid waste incineration.

Title IV. Acid Deposition Program

The stated purpose of Title IV is to reduce the annual emissions of SO_2 by 10 million tons per year, and of oxides of nitrogen by approximately 2 million tons per year from the 1980 levels. The provisions of this title apply to fossil fuel fired electric utilities boilers in specified states.

Title IV implements the SO_2 reductions by utilizing a market allowance system, an imposition of excess emission fees on those exceeding an allowance, and a deadline extension for additional control technologies to reduce SO_2 emissions.

SO_2 Provisions

The SO_2 reductions, beginning in 1995, target emissions from large, higher-emitting utility power plants. These focus primarily on the coal- or oil-fired electric utility facilities. Under this regulation, there will be a national cap on the amount of utility SO_2 emissions in the 22 specified states at approximately 8.9 million tons per year beginning in the year 2000.

The key to the SO_2 reductions is a system of *marketable emission allowances*. In essence, an *allowance* provides an affected source with the authority to emit one ton of SO_2. The affected sources must not emit more sulfur dioxide than they hold allowances for. This market-based system is intended to allow the most efficient utilization of resources and emission credits.

Both spot auctions and advance auctions are provided for on a sliding scale. These allowances can be purchased, sold, and/or banked for use in a later year. All new sources are required to buy into the allowance system. New sources will not be allocated allowances, but will have to purchase them, starting in the Phase II portion of the program.

Market-Based Allowance Program

As noted, allowances will be issued annually to affected utility sources based on their baseline fuel use and the emission rate required by the legislation. Once allocated, an allowance is fully marketable. The fundamental allowance rule is that an affected source must hold enough allowances to cover its emissions.

Sources can comply with this rule through one of several methods:

- by reducing emissions (through the installation of pollution control equipment, fuel switching, or conservation) to the level of allowances it holds; or
- by obtaining additional allowances to cover its emissions.

Sources can also elect to:

- substitute alternative plans
- pool emissions reduction requirements across two or more affected units, or
- craft compliance strategies using limited time extension provisions found in the legislation.

As noted above, a source can comply by obtaining additional allowances sufficient to cover its emissions. These allowances can be obtained in a number of ways:

1. A source can obtain allowances through transfers from other units within its utility system;
2. A source can buy additional allowances on the open market from another source. That other source may have spare allowances for a number of reasons—perhaps because it exceeded its control requirements, thus freeing up allowances for sale permanently or for a limited period of time.
3. A source can obtain allowances from an industrial source that has cost-effective reductions and elects to opt into the allowance system.
4. Sources will be able to obtain allowances through the EPA allowance auctions and sales.

The allowance trading system is expected to provide benefits over the traditional command and control type of regulation:

- First, by not demanding a specific control option, the legislation provides sources with the flexibility to develop the most cost-effective control strategy.
- In addition, the allowance trading system incorporates an incentive for energy conservation and technology innovation, both of which can lower the cost of compliance and yield pollution prevention benefits.

NO_x Provisions

The NO_x reduction program is *not* an allowance-based program. Title IV sets up stringent new NO_x emission factor limits for existing tangentially-fired and dry-bottom wall-fired *coal-fueled* utility boilers. It is expected that low NO_x burner technology will be required to achieve these lower emission rates.

The NO_x emission rates cannot exceed 0.45 pounds/10^6 BTUs for tangentially-fired boilers, or 0.5 pounds/10^6 BTUs for dry-bottom wall-fired boilers and all other burners, unless the EPA determines that these rates cannot be met using low-NO_x burner technology.

By 1997, the EPA must set NO_x emission rates for wet-bottom wall-fired coal-burning units, cyclonic-fired burners, and all other types of utility boilers. The EPA may approve alternative limits if the facility cannot meet the rate limits using low-NO_x burner technology. Emissions averaging between two or more units are possible under the EPA permit program.

Other Provisions

There are provisions for excess SO_2 or NO_x emissions at a penalty of $2,000 per excess ton of emissions, adjusted for inflation. This is in addition to any other liabilities, fees, or forfeitures for noncompliance. In addition, in the case of excess SO_2, the facility must offset the excess emissions by reducing the emissions by an equivalent tonnage the following year.

Large units must comply by 1995, whereas smaller boilers must comply by January 1, 2000. These emission standards will be accomplished by the promulgation of revised *New Source Performance Standards* (NSPS) for NO_x from all fossil fuel-fired steam generating units.

In addition to the emission requirements and/or fee caps and allowances, Title IV requires that emissions be monitored for SO_2, NO_x, opacity, and volumetric flow rate. In addition, "acid rain" permits are required under this title, but will be administratively handled under the provisions of Title V (q.v.). In other words, large utility boilers will have an acid rain permit for Title IV as a part of the *operating permit* under Title V. The initial (phase I) permits will be issued by the EPA, whereas the later (phase II) permits will be issued by states with EPA-approved permit programs.

Control technologies for coal-fired power generation facilities which achieve significant reductions of either SO_2 or NO_x, will be deemed "clean coal technologies." This section of the Act allows for a temporary project to demonstrate clean coal technology. Permanent installations of clean coal technologies will be allowed where there are no emission increases.

Title V—Operating Permits

The goal of Title V is for states to issue federally-enforceable *operating permits* to major stationary sources of air pollution. These operating permits are designed to enhance the ability of the EPA, the state, and citizens to enforce the requirements of the Act. The permitting programs will provide for permit fees to the states to support their programs.

Title V is structured to allow states to develop their permitting program with EPA oversight. This program requires that a federal permit be obtained by each major source in every area within every state, regardless of attainment status. Major sources are defined according to the ozone nonattainment status by geographic area within each state. These also include the 10 ton per year (TPY) sources of a single hazardous air pollutant, or 25 TPY of all combined HAPs.

Other sources coming under the operating permit program include those regulated under any of the following CAAA categories:

- NSPS sources
- NESHAP sources
- PSD sources
- new source review (NSR) sources
- acid precipitation sources (Title IV)

There are some possibilities for exemptions or additions by the EPA. An exemption may be found for a category where the permits would be considered impractical, infeasible, or unnecessarily burdensome. Figure 9.1 illustrates the overlap of the various programs requiring a permit under this Title.

Permit Program Requirements

For each permit to be issued for a facility, it must contain specific provisions which will ensure that the requirements of the permit are being realized. These are:

- fixed term, not to exceed five years,
- permit conditions to assure compliance,
- schedules of compliance, including remedial measures,
- inspection, entry monitoring, compliance certification, and reporting requirements to assure compliance with terms and conditions.

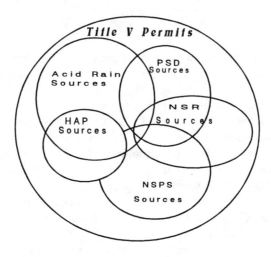

Figure 9.1.   The Clean Air Act permit structure.

In order to be approvable, each permit program under state provisions must contain the elements seen in Table 9.10. As seen, the provisions of the state program and the state permit are extensive. It should be noted that the state or local agencies will issue one federal permit for a facility with multiple sources of air contaminant emissions.

Fees

Each approved state permit program must require the source of the air contaminants to pay an annual fee to the state sufficient to cover "all reasonable costs" associated with developing and administering the permit program. These fees begin at $25 per ton, excluding carbon monoxide. The fee may be increased each year in accordance with the Consumer Price Index.

It should be noted that the state is not required to charge fees for emissions of any pollutant from a single source in excess of 4,000 tons per year. If, however, the EPA determines that a state's fee program is not approvable or the state is not adequately administering or enforcing the fee program, the EPA may step in and collect the fees from the permittee. Failure to pay a fee results in a penalty of 50% of the amount due, plus interest.

Small stationary sources coming under the provisions of Title V as small businesses may incur reduced fees.

Table 9.10.   Title V Operating Permit Program Requirements

Each approved state permit program must contain provisions for:

1. Permit applications and criteria for determining completeness.
2. Monitoring and reporting requirements.
3. A permit fee system.
4. Adequate personnel and funding.
5. Authority to issue permits and assure that each permitted source complies with applicable requirements.
6. Authority to terminate, modify or revoke and reissue permits "for cause" and a requirement to reopen permits in certain circumstances.
7. Authority to enforce: permits, permit fees and the requirement to obtain a permit, including civil penalty authority and "appropriate criminal penalties."
8. Authority to assure that no permit will be issued if EPA objects.
9. Processing applications and public notice, including:
   • offering an opportunity for public comment;
   • a hearing on applications;
   • expeditious review of permit actions; and
   • state court review of the final permit action.
10. Authority and procedures to provide that the permitting authority's failure to act on a permit or renewal shall be treated as a final permit action solely to allow judicial review.
11. Authority and procedures to make available to the public any permit application, compliance plan, permit, emissions or monitoring report, subject to confidentiality provisions.
12. Allowing operational flexibility at the permitted facility.

Other Provisions

A new item required under the operating permits fee program is that the state or local authority is required to notify all contiguous states whose air quality may be affected by the facility receiving the permit. In addition, all states within 50 miles of the application source must be notified.

All states contiguous to the state of the permitted source must be given the opportunity to submit written "recommendations" for the permit to the local or state agency. If the permitting authority refuses to accept those recommendations, it must provide written notice of its reasons to the state which submitted the recommendation, as well as to the EPA.

EPA reserves the right to review the local or state agency permit. EPA has the authority to reject and require revisions to a proposed operating permit from a local or state agency and EPA retains final authority to issue or deny operating permits. EPA may waive its own and neighboring states' review of permits for any category of sources, except major sources, either when approving an individual program or in a regulation applicable to all programs.

Title V also provides for state *court review* of the permit application initiated by anyone who participated in the public comment process or permitting activity. *Any person* may petition the EPA to veto a permit if the EPA fails to object during a 45-day EPA review period. Objections to the petition must have been raised during the comment period. The EPA must issue an objection if the petition demonstrates that the permit is not in accordance with the Clean Air Act.

Valid Permits

If a source complies with its permit conditions, the permit indicates that the source is deemed in compliance with the provisions of the Clean Air Act. However, every three years the permit is subject to reopening and to revision, depending upon the promulgation of other relevant provisions of the law.

Permits issued by the states may be more stringent in their terms and conditions than federal requirements, provided they are not inconsistent with the national permit requirements of the Clean Air Act.

Title VI — Stratospheric Ozone Protection

In order to address concerns for protection of stratospheric ozone, certain compounds containing bromine, chlorine, fluorine, and carbon have been implicated. Title VI of the Clean Air Act Amendments sets up a series of regulations to phase out all emissions of such substances.

The EPA has published two classes (I and II) of substances containing chlorine, fluorine, bromine, and carbon (Table 9.11) which have the potential to deplete stratospheric ozone levels. The Class I substances are composed of bromo- and chloro-fluorocarbons (CFCs); the Class II substances contain hydrochloro-fluorocarbons. The bromine-containing compounds are termed Halons.

For each of these substances, the EPA has established a phase-out schedule by the year 2000 for all CFCs, Halons, and carbon tetrachloride, which are included in Class I. Class II substances, which appear to be less ozone destructive, are scheduled for complete phase-out later. Administratively, the phase-out of CFC manufacturing was moved up to 1995 based upon the federal government's approval of the London and Copenhagen revisions to the Montreal Protocol.

Table 9.11.   Ozone-Depleting Chemicals

Class I Substances
CFC Isomers 11–13, 111–115, 211–217
Halon Isomers 1211, 1301, and 2402
Carbon Tetrachloride
Methyl Chloroform

Class II Substances
HCFC Isomers

Recapture and Recycling

The use and emission of Class I substances are to be at the lowest achievable level. Recapture and recycling are to be maximized. These include air conditioning and refrigeration units at any stationary source. For any source after January 1992, there must be removal and recycling of these substances before disposal.

Venting of either Class I or Class II substances is banned. This Section also sets up a requirement that persons servicing refrigerant-containing equipment must be *certified* to perform the recapture and recycling of the refrigerant during motor vehicle air conditioning services.

CFC-containing small quantity items, such as party streamers, noise horns, cleaning fluids for photooptic and electrical equipment, and other nonessential consumer products are banned. HCFCs (Class II substances) are banned from aerosol containers or plastic foam products, effective January 1994. Foam insulation and rigid foams for auto safety standard requirements are exempt.

Labeling

On all containers of Class I or Class II substances, mandatory warning labels are required. In addition, products manufactured with a Class I substance must also have a warning label on the final product unless substitutes are not available. Class II substances in products or utilized in making products must have a warning label that they contain ozone-depleting substances.

Other Provisions

The EPA must develop requirements for each federal department and agency to conform with its procurement regulations, to maximize the substitute of safe alternatives to ozone-depleting substances.

By 1994, the EPA must produce five reports on methane and its ozone-depleting potential. These reports must identify the sources of both domestic and international methane emission and must analyze the potential for preventing methane emission increases. In addition, this report will identify and evaluate technical options to reduce or stop the growth in methane emissions.

Safe alternatives to Class I and Class II chemicals must be published with a list of substitutes for specific uses. Companies using substitutes for CFCs and HCFCs must provide the agency with any unpublished health and safety studies on those substitutes under this Title.

Title VII – Enforcement Provisions

Title VII modernizes, updates, and increases enforcement provisions of the Act. Title VII enforcement provides a much more severe set of penalties for noncompliance with the Clean Air Act. In addition to violations and fines, the Clean Air Act institutes new *environmental crimes* and raises to a *felony* a knowing violation of any provisions of the act, in addition to penalties for failure to comply with the permit requirements of Title V.

Administrative penalties include up to $25,000 per day per individual violation (to a maximum of $200,000) and up to $5,000 for a field citation for a minor violation of any aspect of the regulations. For civil violations, penalties up to $25,000 per day per violation, with no maximum, are possible. Criminal penalties include up to 15 years in jail and penalties of up to one million dollars.

Clean Air Crimes

By far the most severe penalties follow the creation of new crimes. People who knowingly release hazardous air pollutants and potentially endanger the life of another person may receive up to 15 years imprisonment and fines up to $250,000. For negligent releases of HAPs which endanger another, misdemeanor prosecutions may be pursued.

Knowingly violating a record-keeping provision becomes a *felony*. In addition, convicted persons may make their companies ineligible for any future federal contracts, grants, or loans. The EPA may extend this prohibition to other facilities owned or operated by a convicted person.

Violations of the provisions of a state implementation plan carry a potential *individual fine* of up to $250,000 or maximum imprisonment of five years. Failure to monitor emissions and certify compliance with specific emission limitations may lead to violations. The amount of civil penalties has been increased, and the definition of an operator of a stationary source is now "any person who is senior management personnel or corporate officer."

Compliance monitoring, as provided in earlier Titles, puts the burden on sources of air contaminations to prove compliance once the EPA makes a "prima facie" showing that the violation is likely to have continued. Successive violations of any provision of the Act may lead to doubling of the original penalties imposed.

Sources are required to *certify their compliance* and to report their own violations. No right of appeal is provided from issuance of administrative compliance orders.

Citizens are given the right to sue for civil penalties, and up to $100,000 may be applied for "beneficial mitigation projects which are consistent with this Act and enhance the public health or the environment." If an alleged violation in the past has been repeated, the citizens may sue over these past violations. Monetary awards up to $10,000 are provided to citizens for information leading to penalties on violators.

Criminal liability has been substantially expanded. *Sources* may be held criminally liable for the negligent acts of an employee who faces no liability under the Title VII amendments.

Title VIII — Miscellaneous Provisions

A number of additional provisions have been included under the miscellaneous sections of the Clean Air Act.

Outer Continental Shelf Air Pollution

Sources of air-contaminant emissions from the outer continental shelf (OCS) are subject to emission control limitations, offsets, permitting, monitoring, testing, and reporting in order to meet the requirements of this Act. This applies to all portions of the outer continental shelf, including the shoreline of the Pacific, Arctic, and Atlantic coasts and the Gulf coast of Florida, including vessels servicing or associated with OCS sources.

Sources located within 25 miles of the coast are required to comply with the same requirements as those for sources located in the adjacent shore area. Some exemptions are available based upon feasibility or health and safety concerns, but any resulting excess emissions from such a source must be offset.

Other provisions require that a report to Congress be presented evaluating the health and environmental impacts from combustion of contaminated, used fuel oil in ships, stating the reasons for using such oil and alternatives to such use, along with the associated costs of such alternatives.

Visibility and Source Receptor Concepts

The Act provides for the establishment of *visibility transport regions* and corresponding *Visibility Transport Commissions*. These refer to the interstate transport of air pollutants into Class I PSD areas. The purpose of this commission is to evaluate all information regarding *visibility* impacts and recommend policies and strategies for addressing regional *haze*. Recommendations address how air quality is to be protected through clean air corridors, areas where additional restrictions on emissions may apply, and addresses potential regional regulations for visibility impairment.

Programs sponsored by the EPA and the park service, as well as other federal agencies, are conducted to provide a foundation for developing visibility control programs. The impacts of all other provisions of the Act are to be assessed for their impact on visibility reduction and haze.

A Grand Canyon visibility transport region with a commissioner, patterned after the ozone transport commission required under Title I of the Act, is established by the EPA. Reports on source-receptor relationships for visibility are assessed in reports to Congress.

Other provisions of Title VIII require that the EPA, in conjunction with NASA and the Department of Energy, conduct a study and test program on the development of a hydrogen fuel cell electric vehicle. The study and test program determines how to transfer existing NASA hydrogen fuel cell technology into a mass-producible, cost-effective hydrogen fuel cell vehicle. The program includes a feasibility design study, construction of a prototype, and a demonstration. The results and extrapolation of these tests are expected to produce significant opportunities for low or zero emission vehicles in nonattainment areas.

Title IX – Clean Air Research

Following ten years of the National Acid Precipitation and Assessment Program (NAPAP), this title provides for a continuation of that research effort and the establishment of a task force which will identify any significant research gaps of that previous study and establish a coordinated program to address current and future research priorities. In addition, the Acid Precipitation Task Force coordinates and sponsors additional research, and publishes and maintains a national acid lakes registry. Biannually, the task force submits a report to Congress describing the results of its investigations and analyses.

Due to the lack of verified modeling approaches for catastrophic releases of hazardous air pollutants, Title IX institutes funding for an experimental and analytical research effort dealing with spills of liquefied gaseous fuels. This funding provides for additional chemicals to be included in field testing of releases at the liquefied gaseous fuel spill test facility. The emphasis is on those chemicals that present the greatest potential health risk to human beings as the result of an accidental release. The purpose of the research is to better understand atmospheric dispersion of such releases, and to evaluate the effectiveness of hazard mitigation and emergency response technology.

In addition to the above tasks, the EPA is conducting a research program to identify, characterize, and predict air emissions related to the production, distribution, storage, and use of clean alternative fuels. The purpose of this program is to determine the risks and benefits to human health and the environment, relative to the risks and benefits derived from using conventional gasoline and diesel fuels.

Other research efforts include evaluations of international air pollution control technologies, the effects of acid deposition on lakes in the Adirondack mountains, western states acid deposition, including episodic acidification, particularly at high elevation watersheds west of the Mississippi River.

Titles X and XI

Title X deals with disadvantaged businesses and requires that not less than 10% of total federal funds for any EPA-funded research related to the Clean Air Act Amendments will be available to disadvantaged business concerns. Included among these are universities and colleges with high percentages of minorities.

Title XI provides an amendment to the Job Training Partnership Act, and refers to the Department of Labor. It is therefore a stand-alone Act and does not directly pertain to the Clean Air Act. In this Title, the Secretary of Labor may make grants to states, employers, associations, and representatives of employees to provide training, adjustment assistance, employee services, and needs-related payments to individuals adversely affected by compliance with the other provisions of the Clean Air Act. Fifty million dollars were appropriated for the first fiscal year.

NONREGULATORY AIR QUALITY MANAGEMENT APPROACHES

A number of other approaches may also help to reduce air contaminants, both in the near- and the long-term. These approaches are society-wide, involving government action, industrial action, and consumer awareness.

Source reduction, a strong point in many of the new approaches to *hazardous waste minimization,* has a corresponding benefit in the air quality management field. In general, activities which reduce the generation of hazardous waste also contribute to decreases in fugitive air contaminant emissions.

On the industrial side, one of the programs voluntarily agreed to by over 300 major industrial firms has been the "33/50" program. In this effort, industrial manufacturing firms agreed to a 33% reduction in *hazardous waste* by the end of 1992, with a 50% reduction by 1995. Corresponding decreases were greater for air contaminant emissions, typically of VOC and hazardous organic air pollutants. Other sectors of society which have been contacted by the EPA are the transportation and energy sectors with a similar voluntary program.

Another voluntary program proposed by the EPA pushes for greater energy efficiency. One of these is the "Green Lights" program in which reductions in the wattage of incandescent light bulbs has been proposed. Substitution of lower wattage alternate light bulbs has also been promoted. The presumption is that a reduction in energy usage has a corresponding decrease in criteria air pollutant emissions due to reductions in fuel combustion for power generation.

In the long term, the EPA has been promoting their environmental education curriculum which incorporates pollution prevention throughout secondary schools (K through 12). This is an attempt to heighten the long-term awareness of school-age children to the general aspects of air pollution in their daily lives.

THE INFLUENCE OF NONREGULATORY ACTIONS

There are two major areas which may significantly impact air quality management strategies and, in particular, compliance. The first of these are nonregulatory documents or activities. The second deals with court decisions.

The influence of nonregulatory documents or activities cannot be ignored. These include policies, administrative or executive orders, guideline documents, and standard test methods. *Policies* of a regulatory body set priorities and internal preferred strategies for regulatory action which are not spelled out by regulations. These are discretionary decisions which may have significant impact on individuals, localities, or sources of emissions.

Administrative or executive orders tend to focus on either organization or priorities for resources within an organization. Typically, an air quality management agency has limited resources and, therefore, must prioritize those which are available for dealing with the multitude of laws, regulations, enforcement, and monitoring programs currently in effect. These orders may significantly impact or delay action by regulatory bodies, such as action concerning a permit.

Guideline documents are typically written reports of approaches, favored techniques, or models which, while not legally required to be followed, form the basis of most procedures in determining compliance. Even when incorporated by reference into laws or regulations, these guideline documents may be changed without changes in either the law or the regulations themselves. Thus, they have the force and effect of law without going through the legislative process.

Methods of testing and analysis may have significant impact on compliance and/or air quality management strategies. Inventories of relative emission rates of different contaminants are highly dependent upon the test methods which are used in formulating those inventories. Thus, if a test method is in error or does not accurately represent the true day-to-day emissions of various sources of air contaminants, significant errors in the inventories may result. This would lead to an inappropriate emphasis on one strategy which might be virtually ineffective in actually lowering pollutant concentrations. Thus, an appropriate test method may be at the heart of choosing effective strategies to attain the ambient air quality standards.

Accuracy of models (such as the Urban Airshed Model) which are used as tools in evaluating strategies may have significant impacts on the emission reduction regulations being adopted.

Court Decisions

Decisions by courts function as clarifications for either laws or regulations. Court decisions are typically based upon a point of law which is brought to bear on a deficiency in a regulation or law. Or, they may clarify apparently conflicting laws as to which has priority. The air quality management field is replete with lawsuits in which court decisions are used to force either agencies or sources of air-contaminant emissions to address a particular problem which is a focus of the lawsuit.

Lawsuits filed by environmental groups in the past have forced the EPA, for instance, to list chemicals for regulation. In general, court decisions tend to clarify and/or redirect priorities for regulatory bodies. Citizen lawsuits also form a large portion of court decisions, and many cases are "taken on" as representative of the public interest.

10 MANAGEMENT, TRENDS, AND INDOOR AIR QUALITY

*Let us a little permit Nature to take her own way; she better
understands her own affairs than we.*
                                        Montaigne, *Essays, bk III*, 1595

In the field of air quality management, society's primary task is to come
to a true understanding and definition of the problem. This includes an
understanding of the trends seen in the past, those perceived for the future,
and the relative contribution of air quality levels to our personal environment. Each of these is interrelated in the big picture of air quality resources
and their management.

The scale of the issues ranges from individual health, based upon ambient
air quality standards, to the potential for vegetation effects at the regional
or continental scale. Global issues, such as the potential for climate change
are worldwide in scope.

Other considerations which must be taken into account are the trends in
effectiveness which we have seen through previous efforts at managing air
quality resources. From these, society is better able to understand the success or failure of approaches taken in the past.

From these efforts, we come to an understanding of the economic costs
of implementing such strategies. All resources are limited to some degree,
and therefore society must balance a number of factors, among which are:
human health, the costs to provide different levels of control, the naturally
occurring background contaminant concentrations, the relative importance
of anthropogenic versus natural sources, and even the potential for adaptation of various species to different levels of air quality.

One of the areas of recent concern in urbanized areas is indoor air quality. There are significantly different pollutant gases and concentrations indoors than outdoors. Therefore, the approaches taken for protecting indoor air quality must be of a different kind than those taken for the exterior environment. Likewise, the issues of personal preference versus personal freedom enter the picture.

AIR QUALITY MANAGEMENT

Management as a technical discipline has been described as one in which five distinctive steps are continuously being performed. These include:

- definition
- planning
- implementation
- control
- evaluation

With the management of any operation or resource, the first step must be a full *definition* of both the *objective* and the *problems* that appear to interfere with attainment of that objective. The next step in management is to *plan* the approach: how and what is to be accomplished in the management scheme.

The *implementation* of the planned approach requires the execution of each of these steps and the enforcement of such steps. *Control* of the activities involved which lead to the attainment of the defined objective is next. The final step in the process is an *evaluation* of the effectiveness of each of those previous steps.

This process of management then returns to a redefinition of the problem and the objectives. These include refined planning based upon the findings of the evaluation, and whether the objectives have been redefined or remain the same. Ultimately, this loop continues until the final attainment of the objective.

In the field of air quality management, there are strong correlations between general management and the approaches taken in *air quality management*. Figure 10.1 is the air quality analog of a typical management approach.

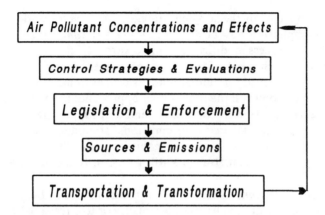

Figure 10.1.   Elements of air quality management.

Elements of Air Quality Management

A *definition* of the effects of air pollutants on the general population, and an assessment of the objective — minimal levels of air contaminants — is our starting point. With respect to concentrations and effects, we have seen earlier that there is general agreement on human health impacts for criteria air pollutants These are verified in the setting of air quality standards. Noncriteria air contaminants have much less data available, and therefore the evaluation of effects and acceptable levels is still somewhat unclear (beyond risk reduction approaches).

This leads to the second step in the process — an evaluation of *strategies* which may be most effective in attaining the desired goal. These evaluations consist of analyzing the trends monitored in ambient air and comparing them to the approaches adopted in the past. Ongoing research in atmospheric chemistry, investigations of emissions control improvements, and different approaches to emission reduction are factors included in these evaluations.

The next step is to implement the control strategies by legislation and enforcement action. This step impacts sources and their emissions. *Implementation* of selected control strategies by legislation and enforcement proceeds from an understanding of air pollution effects and control strategies. In the United States, the Clean Air Act Amendments of 1990 set a fairly definitive implementation strategy for the attainment of ambient air quality standards.

The sources of emissions reflect the scale and the sectors of society which are most impacted by the adopted strategies. As seen for stationary sources, a greater emphasis is now being placed upon *pollutant prevention,* while the emphasis on process modifications and control technologies has continued. Indirect approaches have focused on techniques to reduce the number of vehicle miles traveled for automobiles, as well as tighter emission standards for each automobile on the road. Other approaches have taken the form of encouraging mass transit, controlling land use, supporting car pooling programs, and additional taxes on emissions of undesirable contaminants. Changes in future power plants for mobile sources will be one area in which air quality management strategies will focus on feasibility and technology research.

Transport and transformation of air contaminants in the air lead to the attainment of the desired levels of air quality or to a reevaluation of further control strategies. If the latter is indicated, we again "enter the cycle." The transportation and transformation of air contaminants in the atmosphere is an ongoing area of research. A better understanding of dispersion and location effects leads to a better understanding of source impacts at local receptor sites, particularly for hazardous air pollutants.

As our understanding of air pollutant effects is refined, there will be changes in control strategies and a reevaluation of existing strategies for attaining acceptable levels of air quality.

TRENDS IN EMISSIONS

The trends we speak of include past trends in air quality levels, in contaminant emissions, and in management strategies. Based on an assessment of these past trends, an understanding of the principles of air quality management and the steps therein, limited projections of future trends in each area (air quality concentrations, emissions and further controls), are possible.

Significant emission reductions have been accomplished in the last 20 years of air quality management strategies. Figure 10.2 illustrates the reduction in emissions over this two-decade period for the five criteria pollutants and organic compounds (as a surrogate for ozone). The most dramatic success stories have been in the reduction of lead and carbon monoxide emissions over this period.

Over 50% reductions in highway CO emissions have occurred over the past decade, while national vehicle miles have increased by 30% to 35%. This illustrates the impact of automobile control strategies on just CO emissions.

Figure 10.3 illustrates the significant impact in just 10 years of CO emis-

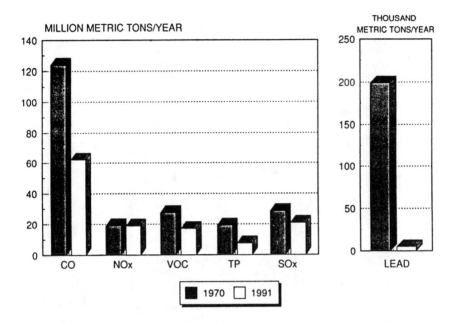

Figure 10.2.   National emission trends.

sions from the transportation sector and in fuels combustion. A significant number of federal air monitoring stations have attained the ambient air quality standards in the last few years. Other locations, as noted earlier, are in nonattainment classifications. A closer look at the emissions of lead is seen in Figure 10.4. Over the last 10-year period, in excess of 90% of lead emissions have been eliminated. The majority of these have been in the transportation source category. The ambient air quality standard is being attained by the overwhelming majority of all federal monitoring sites in the country.

Oxides of nitrogen emission trends (Figure 10.5) over the last decade have shown only slight reductions. These have occurred primarily as a result of improvements in transportation emissions. The majority of ambient air quality monitoring stations are reporting *attainment* status nationwide.

Average air quality for sulfur dioxide, in general, shows attainment for the overwhelming majority of ambient air quality monitoring sites. Reductions in SO_2 emissions (Figure 10.6) have been slight over the 10-year period and have been predominantly in the fuel combustion areas.

PM10 emissions (Figure 10.7) show only a slight improvement over three years, due primarily to reductions in PM10 from fuel combustion. A signif-

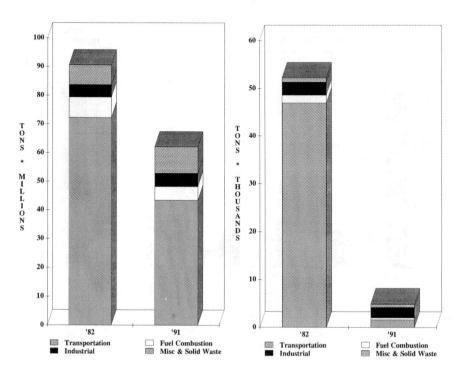

**Figure 10.3.   CO emissions trend,   Figure 10.4.   Pb emissions trend,
            (1982 vs. 1991).   (1982 vs. 1991).**

icant number of reporting stations for PM10 are reporting nonattainment status; therefore, stricter provisions of the Clean Air Act for nonattainment areas may be expected until it improves.

VOC emissions are seen in Figure 10.8 as a surrogate for ozone. The ratio of VOCs to NO_x is probably the more critical component to be evaluated for ozone. As a result of smaller reductions in VOC emissions and nominal reductions in NO_x emissions, it is not unexpected that average ozone levels have changed little over the last decade.

Ozone Trends in the Extreme Area

With respect to ozone, the most significant trends are those in the most extreme nonattainment area in the United States, the Los Angeles basin. It does have the longest history of both monitoring ozone and implementing

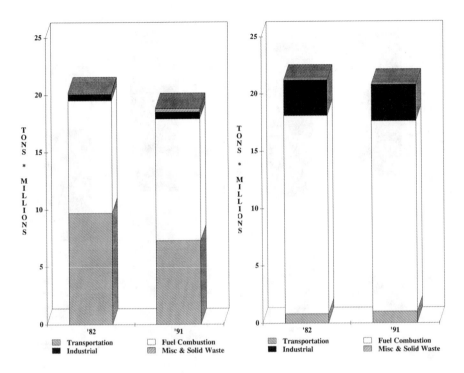

Figure 10.5.   NO$_x$ emissions trend, (1982 vs. 1991).   **Figure 10.6.   SO$_x$ emissions trend, (1982 vs. 1991).**

source control strategies. In Figure 10.9 are the monitored reductions in ozone over 30 years in southern California at three sites which exceeded the federal standard. Significant in this figure is the *decrease in the number of days exceeding the federal AAQS.*

Los Angeles city achieved dramatic reductions in going from more than 150 days per year to less than 40 days per year. Other locations, being closer to the ocean, had significantly fewer nonattainment days. This figure points out the dramatic reductions due to rigorous air quality management approaches in the face of increasing population.

Probably the most significant impact has been the reduction in the total number of days, basin-wide, of high ozone (federal alert) levels over the last 15 years. In Figure 10.10 are the number of days that federal alert (0.20 ppm) and second stage episodes (0.35 ppm) occurred during any single day. Since 1988 there have been *no* second stage episodes. Federal alert levels have been reduced from over 120 days per year in 1977 to 41 by 1992.

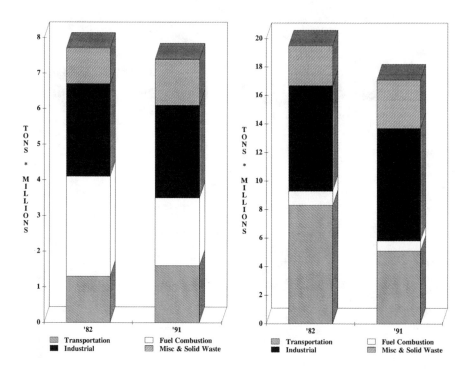

Figure 10.7. PM10 emissions trend, (1988 vs. 1991).

Figure 10.8. VOC emissions trend, (1982 vs. 1991).

These reductions in elevated ozone readings are a result of aggressive implementation of air quality management strategies in southern California. In this most extreme case of ozone nonattainment, a "peak shaving" phenomenon has occurred in which significantly fewer hours and days are experiencing severe ozone levels. Likewise, fewer days of nonattainment of the federal ambient standard are occurring. However, it appears that within the southern California area, some areas will continue to experience days of poor air quality for the foreseeable future.

TRENDS IN STRATEGIES

As a result of the emissions and ambient air quality patterns seen to date, a number of future trends may be inferred. The most extreme case, of

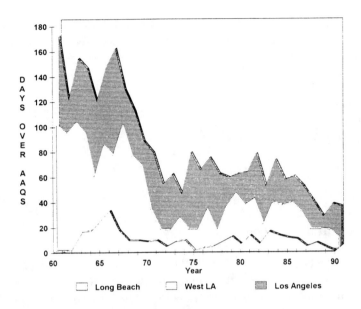

Figure 10.9.   Long-term ozone reductions, Southern California.

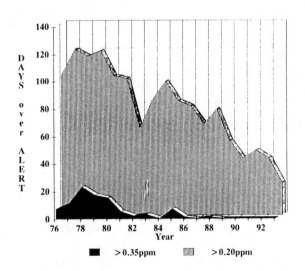

Figure 10.10.   Reduction in high ozone days, Southern California, 1976-1992.

course, is seen in southern California, which is the "test case" for many air quality strategies. Seven general categories are seen as the focus of these management trends:

- fuels and transportation
- small sources
- governance
- hazardous air pollutants
- life style impacts
- stationary sources
- natural sources

Fuels and Transportation

Across the board, cleaner fuels will be pushed by all levels of governmental authority. With respect to mobile sources, we may expect to see compressed natural gas as a near-term, readily available fuel source, until electrification becomes a greater reality for a large percentage of the commuting work force.

The long-term trend is for all transportation in areas of poor ozone air quality to involve some form of electrical-powered individual vehicles, coupled with significant attempts to provide mass transit for the commuting work force.

For heavy-duty vehicles, we may see a shift expected in the power source away from diesels, with their problems of NO_x and particulates, to possibly combustion turbines firing on ultra-clean fuels. This may significantly impact emissions while maintaining the power required to propel heavy-duty vehicles.

Small Sources

In the more serious nonattainment areas for criteria pollutants, there will be an increasing focus on smaller and smaller sources of air contaminants. With southern California as an example, increasing regulations on consumer products, such as solvent content of cosmetics, perfumes, paints, barbecue lighter fluid, and windshield-washing fluids will be seen.

Small combustion sources, such as hot water heaters, will be required to install natural gas burners utilizing high-efficiency, low NO_x combustion systems.

As the regulatory burden continues to reach out to all segments of society, not only will consumer products be affected, but also small businesses. These small businesses will continue to be held to a strict level of air emissions limitations; however, there will be temporary fee breaks or assistance programs to help small businesses attempt to survive.

Governance

As a result of public pressure, greater "public input" may be expected on all aspects of air pollutant reporting and permitting of stationary sources. As seen in the Clean Air Act Amendments, one person's objection to the issuance of an operating permit may cause the entire permitting process to halt and enter a minimum additional 30-day hearing period before any permit may be granted.

In order to complement enforcement, increasingly taxes will be used as a method of reducing emissions, as well as heavier penalties for persons in responsible charge of air pollutant-emitting sources. Personal liability for emissions will enter the picture.

As seen in the amendments to the Clean Air Act, an increasing "federalization" of air quality management strategies, regulations, planning, and permitting will be seen.

As a result of increasing pressures on budgets for various levels of government, we may expect to see a greater number of "fee-based services" as a part of air quality management strategies. Every service or action required as a result of regulatory action will be reviewed by various arms of government, for which specific *fees* will be assessed. This will include reviews of indirect source plans, performance reviews of risk assessments and dispersion models, reviews and evaluations of self monitoring data, and evaluations of the adequacy and attainment of "toxic hot spots" regulations.

As agencies are overwhelmed with requirements to provide increasing services, there will be attempts to "privatize" certain functions formerly performed exclusively by regulatory personnel. This will take the form of air pollution permit application preparers and processors. The latter will be required to be "certified" under the authority of the local regulatory agencies. These certified private permit processors will then do the work of the regulatory agency, under their supervision for a fee to the source owner.

At the local level, as backlogs increase at regulatory agencies due to increasing regulations, there will be attempts to streamline operations. One will be a requirement that standard equipment, such as hot water heaters, package boilers, and certain coating operations, be certified by the manufacturer as meeting stringent air emission standards throughout the life of the particular piece of equipment.

Hazardous Air Pollutants

As a result of concerns over hazardous air pollutants, there will be an increasing focus on toxic and noncriteria contaminants. They will include both carcinogens (suspected and/or promoters) and noncarcinogens which may contribute to other detrimental health effects. These will require increasing controls, as these emissions are orders of magnitude lower than the criteria contaminant emissions.

As a result of the unknowns involved in noncriteria air contaminant effects, there will be an increasing emphasis on "risks" to health as a result of noncriteria air contaminant emissions. Risk assessments will, therefore, continue to occupy a large part of the concern for allowing emission sources to continue to receive permits.

Life Style Impacts

Increasingly, aspects of individual life-styles will be impacted. The focus will be on curbing or restricting individual mobilities (i.e., cars driven by IC engines), consumer products, and smaller combustion sources. In certain areas, one may expect prohibitions on solid fuels, such as wood stoves and wood burners, unless controlled by either high-temperature thermal systems, precipitators, or other particulate control strategies. Smoking will be banned in all public buildings.

There will be a much greater emphasis on indirect sources. In particular, these regulations will focus on vehicle usage and attempts to reduce not only the total number of miles traveled, but also the number of individual trips. Trip reduction relates to emissions before and after the automobile is driven.

Attempts to reduce the number of miles traveled focus on the emissions during operation. Limits on the numbers of vehicle registrations allowed, additional parking lot usage fees, required delayed starts of workdays, and bans on single occupant vehicles are proposed as backup measures to control mobile source emissions in the extreme nonattainment areas.

Stationary Sources

Increasingly, *employers will be the enforcers* of air quality management regulations. In southern California, employers are held responsible for the transportation modes of employees and their willingness to join car pools, van pools, and other methods of reducing indirect source (auto) emissions.

In the extreme nonattainment areas, solid and liquid fuels will be banned for all stationary sources except for essential public services or emergencies.

There will be increasing attempts to incorporate *air pollution "bubbles"* on stationary sources. These will set a cap on total emissions for each individual source and a cap on the emissions throughout an entire region. These programs will involve both internal and external trading and the issuance of certificates or credits, which may be bought and sold. As sources reduce emissions beyond that required by law, those excess emission reduction credits will become available to other sources which may need "offsets" for its new source increases in emissions.

There will be an increasing emphasis on *self-reporting* of emissions or alternative process parameters which may be correlated to emissions. These will include continuous emission monitors (CEM) for large sources and the installation and monitoring of fuel flow, solvents, and raw materials and paints for lesser sources.

Additional management approaches will be required for stationary sources which will involve certifications of employee training, good operating practices, additional record-keeping and reporting, as well as continued testing of emissions to verify compliance.

As the regulatory burden increases, both business and industry will struggle to meet air pollution codes. This will lead to relocations of certain industries to areas with less severe attainment status. In extreme areas, additional taxes of $5000 per ton of VOCs are proposed. In southern California, businesses within 10–12 miles of the coast may be required to open for business after 12 noon as an ozone control measure.

Some businesses, however, will find increasing opportunity. These will be the manufacturers of air pollution control equipment, those providing consulting services, and those providing measurement and analysis services to affected stationary sources. Figure 10.11 is one company's estimate of the potential market for these services by type of industry, between now and the turn of the century.

Natural Sources

Even nature will not be exempt. PM10 emissions from wind erosion are considered sources subject to human control. Control actions will be required from private land owners. Air quality management plans in southern California have suggested maximum speeds of 15 miles per hour on improved public and private roads to be appropriate for PM10 emissions in arid nonattainment areas.

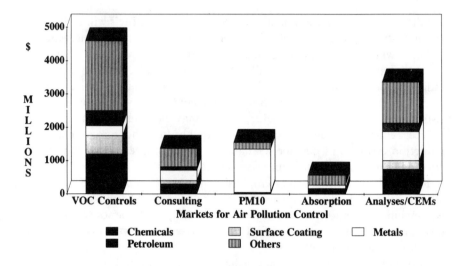

Figure 10.11.   Market estimates for air pollution services.

The Outlook

Overall air pollutant concentrations will continue to drop; however, it may be expected that these concentrations will approach some "asymptotic" level with the implementation of these strategies.

Highly urbanized areas in locations with low inversions, coupled with existing background concentrations of NO_x and high natural sources of reactive hydrocarbons, such as terpines, may find it nearly impossible to reach the federal ozone standards. At that point, other strategies may be implemented.

Upgrading buildings and indoor air quality to counterbalance the time spent in the outdoor environment where the air quality still may not achieve federal standards may come to the fore. This brings us to the next concern: indoor air quality.

INDOOR AIR QUALITY

At the personal level, there is significant concern for the quality of the air inside both residences and public buildings. In Figure 10.12 is seen the percent of a person's day spent inside buildings, as opposed to that time spent outside or in transit for the average person in the United States.

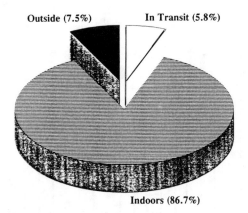

Figure 10.12.   Exposure locations, time percentages.

Remaining indoors during high air pollutant episodes is one method which allows a person to breath more easily with a modest amount of lifestyle interruption. Regulatory agencies have been recommending for decades that people remain indoors to avoid serious air pollutant episode levels. However, this advice is increasingly being questioned due to new information on the levels of air contaminants found indoors. Table 10.1 lists common contaminants of concern with respect to the indoor environment.

With respect to reactive gases like sulfur dioxide and ozone, the advice to remain indoors on a "smoggy day" is well taken. These reactive gases have short half-lives in an indoor environment. Figure 10.13 shows the relative concentrations of SO_2 measured simultaneously indoors and in the outside ambient environment for eight different sites in the Kansas City area.

Ozone, in the absence of high voltage sources, such as photocopiers and/

Table 10.1.   Indoor Air Pollutants of Concern

- Criteria Pollutants
- Environmental Tobacco Smoke
- Radon
- Combustion Emissions:
    CO, NO_2, Particles, Polycyclic Aromatic Hydrocarbons
- Formaldehyde
- Volatile Organic Compounds
- Pesticides
- Bioaerosols, e.g., molds, spores, infectious agents

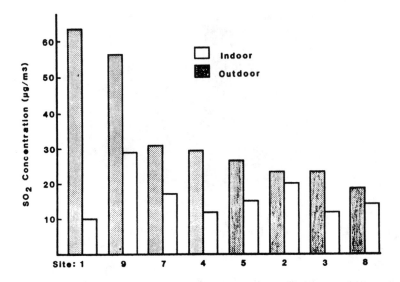

Figure 10.13. SO₂ concentrations measured simultaneously in the indoor and ambient environments.

or electrostatic precipitator equipment, shows similar reductions in indoor air quality concentrations as opposed to those outdoors.

Other gases, such as NO_2, may exhibit appreciably higher concentrations indoors than outdoors due to the mode of generation. In Figure 10.14 are seen the levels of NO_2 at various points in a residence compared to those concentrations simultaneously measured outside of the structure. Depending on location, the indoor NO_2 concentration varies from two to three times higher indoors. This residence had a natural gas fired stove and oven combination which was used for all cooking.

Other measurements of noncriteria pollutants, such as organic gases, show significantly higher concentrations indoors as opposed to outdoors, as well. Table 10.2 lists the results of simultaneous testing of a large number of sources indoors as well as outdoors, for different organic gases. The ratios of personal exposure indoors to those levels found outside range from a factor of three times higher (with styrene) to over 20 times higher for contaminants such as 1-1-1 trichloroethane (TCA).

The variety of indoor air contaminants of concern range from the obvious, such as tobacco smoke, to less obvious ones such as bio-aerosols, which include molds, spores, and other infectious agents.

Of all of the contaminants to which people may be exposed in the indoor

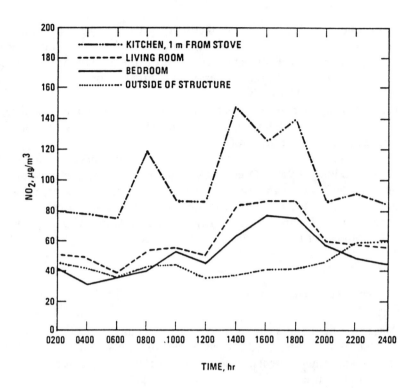

Figure 10.14. Diurnal indoor/outdoor pattern for NO_2 in a home, spring-summer, 1973 (composite day based on 6 days of data).

Table 10.2.   Indoor-Outdoor Relationships – Organic Gases

| | Arithmetic Mean[a] | |
Compound	Ambient Air	Indoors
1,1,1-Trichloroethane	5.4	110
Benzene	8.6	30
Carbon Tetrachloride	1.2	14
Trichloroethylene	2.1	7.3
Styrene	0.9	2.7
m,p-Dichlorobenzene	1.5	56
Xylenes	15.0	71

[a]$\mu g/m^3$

environment, the number one risk factor is tobacco smoke. These include life-style, occupational, and indoor exposures. With respect to lung cancer and early or premature death, smokers who are exposed to radon experience the synergistic effects of both. Other volatile organic compounds, such as formaldehyde, are also on the list.

Table 10.3 summarizes the usual sources of indoor air pollutants. These include combustion, building materials, consumer products, people and, to a lesser degree, infiltration of outdoor air.

Perceived indoor air quality issues common to office workers in buildings are called Building-Related Occupant Complaint Syndrome, or in common terms, *sick-building syndrome*. Not the same but related issues, such as Building-Related Illnesses (BRI) do have clinical symptoms associated with diseases such as Pontiac fever or Legionnaires' disease. These are more properly biological factors than air pollutant factors.

Environmental Tobacco Smoke

That smoking indoors causes indoor air pollutants is a common fact of life. Table 10.4 lists measured concentrations of air contaminants in a variety of smoking environments, as compared to nonsmoking controls in similar indoor structures.

Tobacco smoke is a complex mixture of thousands of volatile, semi-volatile, and particulate organic and inorganic compounds formed from the smoldering combustion of tobacco leaves. In general terms, mainstream smoke is that tobacco smoke which is inhaled directly by the smoker and then exhaled. Sidestream smoke is that which is emitted by the smoldering end of the cigarette between puffs.

Environmental tobacco smoke (ETS) is sidestream tobacco smoke to which nonsmokers are exposed in areas sharing a common air volume. The range of vapor phase constituents from tobacco smoke is listed in Table 10.5 on a mass basis per cigarette. Many of the compounds come under the heading of "tar."

The EPA has estimated that the health effects of environmental tobacco smoke may include up to approximately 3,000 excess lung cancer deaths per year, triggering 8,000 to 26,000 new cases of asthma in previously unaffected children. Symptoms are exacerbated in 400,000 to 1 million asthmatic children per year according to EPA publications.

Studies of nonsmoking office workers exposed to ETS have a demonstrated reduction in small airway function, comparable to those seen in smokers who consume between one and ten cigarettes per day. Studies

Table 10.3.   Sources of Indoor Air Pollutants

Combustion Sources:
- Tobacco Smoke
- Woodburning Stoves
- Kerosene Heaters
- Gas Stoves
- Automobiles in Attached Garages

Building Materials and Furnishings:
- Particleboard
- Paints
- Adhesives
- Caulks
- Carpeting

Consumer Products:
- Pesticides
- Cleaning Materials
- Waxes, Polishes
- Cosmetics, Personal Products
- Hobby Materials

People

Outdoor Air
- Ozone
- CO

Table 10.4.   Tobacco-Related Contaminant Levels in Indoor Spaces

Contaminant	Type of Environment	Levels		Nonsmoking Controls	
CO	Room (18 smokers)	50	ppm	0.0	ppm
	15 Restaurants	4	ppm	2.5	ppm
	Arena (11,806 people)	9	ppm	3.0	ppm
PM	Bar and grill	589	$\mu g/m^3$	63	$\mu g/m^3$
	Bingo hall	1140	$\mu g/m^3$	63	$\mu g/m^3$
	Fast food restaurant	109	$\mu g/m^3$	24	$\mu g/m^3$
NO_2	Restaurant	63	ppb	50	ppb
	Bar	21	ppb	48	ppb
Nicotine	Room (18 smokers)	500	$\mu g/m^3$	—	
	Restaurant	5.2	$\mu g/m^3$	—	
Benzo-α-pyrene	Arena	9.9	ng/m^3	0.69	ng/m^3

Table 10.5.   Some Vapor-Phase Constituents of
Cigarette Smoke

Compound	μg/Cigarette
NO_x	750
CO	58,000
NCN	100
NH_3	9,450
Ethene	2,100
Propene	1,550
1,3-Butadiene	360
Dimethylamine	40
Formaldehyde	2,000
Acetaldehyde	4,700
Acrolein	1,090
Acetone	1,080
Propenal	390
2-Butanone	720
Crotonaldehyde	280
Benzaldehyde	80
N-Pentanal	60
Benzene	650
Toluene	1,260
Styrene	100
N-Nitrosodimethylamine	0.9
Nicotine	1,230
Formic Acid	525
Acetic Acid	1,500
Methyl Chloride	940
Pyridine	370
Phenol	230
Catechol	170
Aniline	10.8
2-Toluidine	3.0
4-Aminobiphenyl	0.14
Benz(α)anthracene	0.14

conducted in Japan have shown that nonsmoking wives of smoking husbands have a significantly elevated risk of developing lung cancer compared to nonsmoking households.

Other risks associated with environmental tobacco smoke are the production and emission of radioactive decay products when radon is present. Radon daughters, which are particulates, remain suspended in the atmosphere for long periods of time, and provide a vehicle for the transport of radon progeny into the lungs.

The use of chemical fertilizers in growing tobacco are transported through the crop chain due to the use of phosphorus-containing fertilizers. These latter compounds have been associated with radioactive polonium, found to be a cancer-causing agent correlated with tobacco smoke.

Analyses of the particle size of cigarette smoke (0.1 to 0.5μ) indicate that the predominant particle sizes are in the aerodynamic range most likely to cause deposition in lungs.

Radon

The relationship between high radon levels and lung cancer death in uranium and other mines has been known for decades. Only recently have we become aware of its subtle health risks in homes and in other indoor dwellings under normal living conditions. Radon, which emanates from soil, building materials, water, and air, is now understood to have some role in the cause of lung cancer. It is estimated by the EPA that in the United States, from 7,000 to 30,000 lung cancer deaths may occur annually due to indoor radon.

Outdoors, the risk from radon or radon daughters is slight. However, inside a building, the release of radon gas to the outside is blocked, and it accumulates in the dwelling. Even in well-ventilated dwellings, the indoor level is higher than outside. With well-constructed and well-insulated homes, radon may accumulate to high levels.

Radon is the heaviest of all naturally occurring inert gases. In the periodical table, it is element number 86. Since it is an inert gas, it is not prone to chemical reactions and therefore, as soon as it is produced, it is able to move freely through pores in the soil and cracks through walls. Radon is produced by the decay of radium, another element which occurs naturally in rocks and soil. Since radon is also radioactive, it will decay into another element as soon as it is produced, eventually ending up as a stable element, lead. Figure 10.15 illustrates this decay sequence.

Radon gas is soluble in cold water; however, its solubility decreases with increases in temperature. Since the gas is constantly being produced naturally from rocks and soil, all homes and buildings contain some radon.

The basic unit of radioactivity, set by the International Radium Standard Commission used only for radium and its decay products, is the Curie (Ci). One Curie equals 37 billion (or 3.7×10^{10}) disintegrations per second of radioactivity, which was arbitrarily set for general use. The magnitude of the Curie was historically based on the activity of radium, and is approxi-

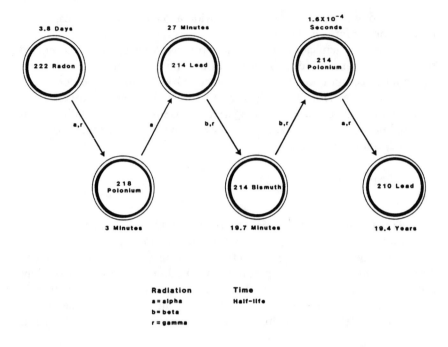

Figure 10.15.   Radon radioactive decay chain.

mately equal to the rate of decay of one gram of Ra-226 per second. However, as the accuracy of measurements increased, the exact value has changed. Since the radioactivity of radon is very small, it is usually reported in *picocuries* (pCi), or 10^{-12} Curies. The amount of radioactivity of radon is reported per liter of air.

The EPA regards 4 pCi/l as the action level for indoor radon and recommends remedial action if the level is higher. Other agencies recommend remedial action at different levels. The International Council on Radiation Protection (ICRP) recommends remedial action at 8 pCi/l or higher. The World Health Organization (WHO) standard is 10 pCi/l for existing buildings, but 3 pCi/l for new buildings. In Canada, the standard is 2 pCi/l. These standards are arbitrary.

The EPA standard is a statistical one and is not based on a threshold health effect. The 4 pCi/l level is the rounded-off geometric mean value of homes tested in the United States, plus one standard deviation (Figure 10.16).

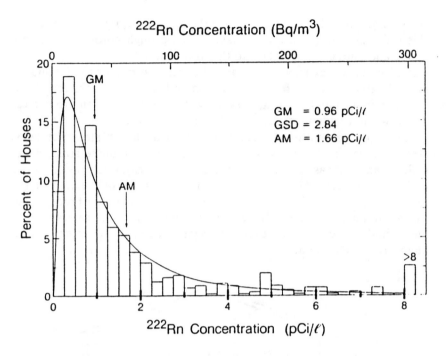

Figure 10.16.   USEPA distribution of radon concentrations in residences.

Radon in Building Sources

Soil

Radon in soil gas is affected by several factors such as radium content and distribution, the porosity of the soil, moisture content, density and the material types, and the underlying bedrock. The radium content of soils reflects the different types of parent bedrock from which the soils were derived. Rock types with high uranium concentration are granites, metamorphosed igneous rocks, and shales.

Soil-gas radon levels vary widely even in the small areas, such as within a housing tract, and are not well-correlated with radium content of the soil. Soil *porosity* has a high correlation with soil-gas radon.

High permeability soils in combination with very small pressure differentials in buildings increase indoor radon levels easily. In soils such as sand

and gravel, radon diffusion is high. On the other hand, relatively imperme-
able soils, such as clays, do not have a high enough porosity to allow a
significant amount of soil gas into the structure.

Fractured bedrock may also be an important contributor to high indoor
radon levels. These are from the sheared and fractured rocks associated
with mountain building. Fractures allow transport of radon from great
depth, where uranium can be located along the fractures.

The major mechanism of radon transport through soil into structures is
pressure differential. The stack effect (Figure 10.17) is due to atmosphere,
building and soil gas pressure differentials. It is a major contributor to air
infiltration, especially in winter. Atmospheric pressure differentials of 1%
to 2%, due to warm/cold front passage, produce inversely proportional
changes in the radon levels. Increased atmospheric pressure forces air into
the ground, while lower atmospheric pressure draws the air together with
radon from the ground.

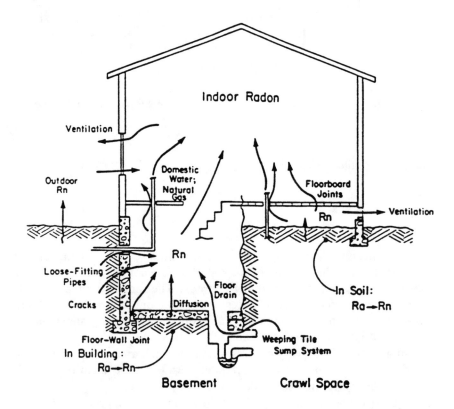

Figure 10.17. Building stack effect.

Water

In the United States, about half of the population relies on underground water, while the other half relies on surface water, rivers, or lakes for water supplies. Among those who use underground waters, most use public supplies which are treated and distributed. Those who use underground water drawn directly from private wells or from community wells may have the water supply as the major source of indoor radon.

When the groundwater is exposed to the surface air, the radon content will be reduced drastically, within days, to a low level, typically around 10 pCi/l. Also, when the water is heated or agitated, radon in water is reduced dramatically.

Thirty to seventy percent of radon will be released from the water just by running water out of a faucet. Over 90% of radon will be released from water when a dishwasher or clothes washer is used. This can be a significant indoor radon problem for families which use high radon content groundwater directly from the well. However, for most people, water is not a source of indoor radon problems.

Building Materials

Although concrete is one of the strongest emitters of radon of all materials, most have relatively low source strengths. The use of phosphate slag in concrete in the United States has been studied, and it has been estimated that there are as many as 74,000 homes built with concrete containing up to 19,980 pCi/l of Ra-226. Bricks tend to have slightly higher Ra-226 concentrations, but have significantly lower radon emission rates than concrete. Red bricks are thought to have higher uranium content than lighter-colored bricks. However, since they are mostly used in the exterior of the buildings, they are of less concern.

Other building materials, such as granite, marble, sandstone, and volcanic rocks, have a wide range of Ra-226 concentrations and radon emanation rates. Overall, most building materials in the majority of homes in the U.S. do not have significant problems with radon.

Measurement and Action Levels

EPA publications contain guidelines for radon and radon progeny sampling procedures. They provide guidance for using seven different types of instruments. For radon measurements, they are (1) charcoal canister, (2) alpha-track detector, (3) grab sampling, and (4) continuous radon monitor. For radon daughter measurements, continuous working level monitors, radon progeny integrated sampling units, and grab sampling are available.

If the radon level is found to be less than 4 pCi/l, no action is required. However, a follow-up measurement is considered a good idea. Although there may be some risk of lung cancer, reduction from this low level is not easy.

If the measured radon level is between 4 and 20 pCi/l, the EPA recommends follow-up tests to determine the average levels for each season of the year. If follow-up tests indicate a similar level, the EPA recommends the level be reduced to less than 4 pCi/l within a few years. If the level is between 20 to 200 pCi/l, a follow-up test within three months is recommended. If the level is higher than 200 pCi/l, the EPA recommends immediate remedial action. Follow-up measurements not more than a week following remediation are recommended.

House Characteristics and Radon Gas Entry

There are many types of dwelling structures. Some are constructed on grade with soil, while others are either subgrade or above grade. Basements are subgrade, while wood houses are constructed above grade over a foundation. The space between the grade surface and the bottom of the floor is the crawl space. Homes on grade have a cement slab under the floor.

The basements are most vulnerable to radon entry. There are usually cracks between the floor and the walls of the basement. Additionally, there are numerous holes and cracks in the walls and spaces around pipes and wiring conduits. Due to the "stack effect," radon gases can enter dwellings easily.

Those built over a cement slab are usually built over sand or gravel that may or may not be sealed. Sealants are commonly used to prevent moisture from entering the cement and are also effective in preventing radon entry. However, cracks can form in slabs, around pipes, and other openings that are not sealed tight enough.

Mitigation

In all mitigation techniques, the goal is to reduce the indoor concentration of radon. Depending on the measured concentration of radon, costs range from inexpensive to costly. If mitigation is necessary, homeowners are usually advised to contact state or local agencies for advice in the selection of a mitigation technique. In some states a contractor's license is required for those who engage in radon mitigation.

If the radon concentration is low, some EPA-recommended measures may suffice: avoid areas of higher radon concentration (that is, spend less time in the basement or areas of the house directly above the soil). Also, stairwell doors, fireplace dampers, and laundry chutes that go to the basement may be closed to reduce radon from lower levels. One should also keep in mind that combustible appliances can cause negative pressurization inside the dwelling.

The most effective technique is to replace indoor air with outside air. In moderate climates, ventilation is perhaps the least costly method. Increasing the ventilation of the living spaces by opening windows is usually recommended. Crawl space vents, if so equipped, may be opened to the outside of the structure.

Subslab ventilation (Figure 10.18) is the current recommended "mechanical fix." Here, clean outside air is drawn into the sand/subslab materials

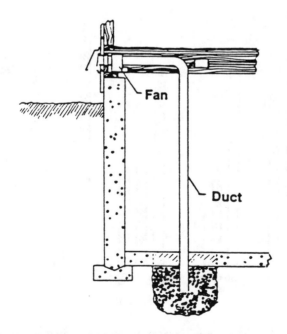

Figure 10.18.   Schematic diagram of subslab ventilation system.

which vents the soil gas away from the structure, thus reducing the infiltration rate.

Visible holes, including the exposed soil in the basement, are sealed off. The exposed soil in the crawl space is also sealed. Other holes and cracks to look for and seal include uncapped sump pits, floor drains, perimeter drains, utility penetrations, and cracks in walls and floors. Cracks and pores on block walls may not be easily spotted.

Airborne particles can be removed by fans, filters, or electrostatic precipitation. However, the effectiveness of such a technique is not certain at this time. Figure 10.19 illustrates the relative effectiveness of each in reducing

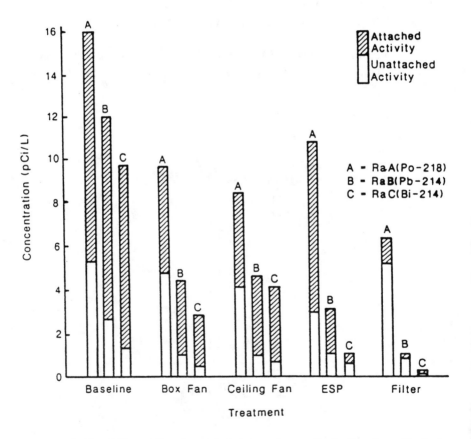

Figure 10.19.   Effect of various air treatments on attached and unattached radon decay product levels.

the gas and particulate phase radioactivity for the polonium, bismuth, and lead isotopes resulting from radon decay. Of course, with any collector, it will eventually become a *source* following usage. Care must therefore be taken for disposal of the collected media.

In colder climates, where opening windows for ventilation is not feasible, forced air ventilation with heat recovery may be cost-effective. The process works best when the indoor-outdoor temperature differentials are large. The indoor air is drawn through a heat-recovery ventilator where the heat of the indoor air is transferred to the cooler outdoor air. Heated outdoor air is then supplied into living areas through ducting, while the stale air is blown out.

OTHER INDOOR AIR CONTAMINANT CONCERNS

Formaldehyde, a listed carcinogen, while it may be much lower on the list of hazards, nevertheless is present in a significant number of residences. Sources of formaldehyde include pressed wood products, such as plywood, particle board, and fiberboard, which are made using urea formaldehyde resins.

These compounds are chemically unstable, and in the presence of moisture, they undergo hydrolysis to liberate gaseous formaldehyde. A number of symptoms have been associated with this such as eye irritation. Emission rates of these components range from 10 to 15 mg/m^2 per day for new sources of pressed wood products. In addition, free formaldehyde trapped in the resin may be volatilized as a function of heat and time. It is the release of unreacted formaldehyde that is primarily responsible for the high initial formaldehyde levels associated with these emission sources. Mobile homes, with a significant percentage of these products, represent the largest single category of high indoor formaldehyde buildings.

Urea formaldehyde insulation products represent the next largest category of sources, with an emission rate of 15 mg/m^2 per day. Other sources include furnishings, fabrics, paper plates and cups, carpets, and clothing with much lower emission rates of formaldehyde. Figure 10.20 illustrates the range of concentrations of formaldehyde in various buildings and compares them to the outdoor air and the action level in the state of California. The 0.1 ppm action level is the recommended exposure by the National Institute for Occupational Safety and Health (NIOSH).

Other contaminants and their sources which may contribute to indoor air quality relate to poor combustion. The combustion contaminants, carbon

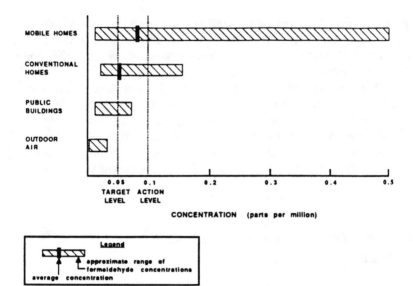

Figure 10.20.   Formaldehyde concentrations in California. Adapted from
Formaldehyde in the Home, Supplement to Indoor Air
Quality Guideline No. 1, **State of California, Air Resources**
Board, Research Division, September, 1991.

monoxide, nitric oxide, PM10, aldehydes and hydrocarbons, are readily generated from stoves, heaters, gas appliances, and wood-burning equipment or fireplaces operating under conditions of poor combustion.

Ozone may build up in confined spaces, typically associated with photocopy machines or in locations where attempts are made to clean the air by use of "ion generators." These latter, due to their high voltages, will build up concentrations of ozone, particularly if they are in confined spaces.

Asbestos, occurring in acoustical ceilings, tiles, and insulation, is the focus of significant regulatory action for school children, as well as some public buildings. The key element in asbestos exposures is whether the material has been disturbed and whether fibers are being released. In many cases, an optimum solution to the concern for asbestos is to ensure that asbestos-containing materials are not disturbed. Clean-out, removal, and remediation of asbestos-containing materials are the subject of an extensive body of regulatory requirements.

The known health effects of both *lead* and *mercury* are sufficient to warrant concerns for older buildings and structures containing lead and mercury in paints. These materials may represent significant hazards, should they be in a poor state of surface condition, causing sloughing of paint fragments into the atmosphere, or where exposure to children may occur where these materials are deteriorating. Again, the removal of these materials or stabilizing them in place, represents a significant investment in lowering exposures to these contaminants.

It should be remembered that there are no current residential air quality standards in existence in the United States. However, many building owners have expressed interest in voluntary "cleanups" of hazards for buildings to which the public has access. Indoor air quality in the workplace may come under federal authority due to OSHA regulations.

In general, for most indoor air contaminants, the approach taken to lessen public exposures and reduce human health effects has been primarily under the category of source reduction. This includes elimination of smoking, either by local ordinance or as a building management requirement. Increasing the ventilation rate, or the number of air exchanges per hour in buildings, has provided significant reductions (up to 90%) in indoor air contaminants.

Improving or eliminating poor combustion will significantly improve the quality of the indoor air. This will reduce the combustion contaminants, as well as eliminate particulates containing toxic compounds.

Elimination of certain solvents, paints, and chemicals and/or replacement of products containing formaldehyde may be considered a significant source reduction approach which will eliminate air contaminants. Building "bake-outs" in which the temperature of the building when unoccupied is artificially raised to high temperatures while increasing the air exchange rates to the maximum feasible have not been found to be particularly effective.

PUBLIC BUILDINGS

There are a variety of indoor building-related occupant complaint syndromes which come under the heading of *indoor air pollution* or poor indoor air quality. In some cases, these have been lumped under the term *sick building syndrome*.

There are distinctions, however. *Building-related illness* (BRI) is a known circumstance where occupants possess clinical disease symptoms. This has been found to be closely associated with bacterial infections as a result of organisms growing in, or in association with, heat exchangers and condensers used for building air conditioning and ventilation.

Table 10.6.   California Healthy Building Pilot Study
Prevalence Odds Ratios[a]

Symptoms	Natural	Ventilation Type Simple Mechanical	Air Conditioned
Runny Nose	1.0	1.7	1.3
Stuffy Nose	1.0	1.9	1.6
Throat Irritation	1.0	2.6	2.5
Sleepiness[b]	1.0	2.0	1.9
Difficulty Breathing[b]	1.0	2.3	2.5
Chills[b]	1.0	4.8	4.9
Skin Irritation[b]	1.0	4.1	3.3

[a]Adjusted in a logistic regression model for gender, job, race, education, age, and smoking.
[b]Adjusted in a logistic regression model only for gender.

The issue of "building sickness" is usually considered an issue which may be more of *perception* than of reality when it comes to common air contaminants. In fact, it is believed that such complaints possess a multifaceted cause. In general, the issue appears to be more occupant comfort than air pollution. Also, it is rare in structures other than office buildings.

In a recent investigation in California where building occupant complaints were registered, 13 office buildings were selected for study. Three had natural ventilation, three had a simple mechanical ventilation (fans), and six were air-conditioned with sealed windows. Data were collected from 880 occupants, along with week-long measurements of indoor and outdoor concentrations of VOCs, CO_2, and carbon monoxide, as well as short-term measurements of fungi and bacterial counts. In addition, temperature and humidity were monitored to determine the results. The symptoms monitored included runny nose, stuffy nose, throat irritation, sleepiness, difficulty in breathing, chills, and skin irritation.

The preliminary results from this California Healthy Building Pilot Study are summarized in Table 10.6. In this table, the prevalence of the symptoms is expressed as the *ratio* of the number of complaints for mechanical (fan) or air-conditioned ventilation systems to a natural (or open-window) structure. These ratios were adjusted in a linear regression model for gender, job, race, education, age, and smoking. This table provides the average ratio as well as the range of prevalence/odds ratios for each specific symptom.

Significantly higher ratios for both the air-conditioned and simple mechanical ventilation were found for chills and/or skin irritation, with lesser prevalence ratios for throat irritation and difficulty in breathing.

Findings of this study found that the higher symptoms were associated with:

- higher job stress
- use of carbonless copy machines
- use of photocopiers
- female gender
- new carpets, and
- no window within 15 feet of the work station

Other findings of this study indicated that the highest levels of CO_2 were found in air-conditioned buildings, along with the least thermal discomfort. The lowest levels of airborne fungi were found for air-conditioned buildings.

Possibly one of the most significant findings was that *no* relationship was found between the symptoms reported and total volatile organic compounds measured in the structures. Thus, the issue of perception versus actual contaminant concentrations appears to dominate the issue. Nevertheless, good management practice would require that even perceived symptoms be alleviated. Better training, higher ventilation rates, and more windows could contribute to lessened complaints, in addition to better humidity and bioaerosol control.

A number of technologies may be adapted from stationary source approaches to control indoor air pollutants where they do occur. These include electrostatic precipitators, scrubbers, activated carbon adsorption units, and high efficiency filters.

BIBLIOGRAPHY

1. THE ATMOSPHERE AND ITS CONTAMINANTS

General References in Air Pollution

1993 Information Please Environmental Almanac, World Resources Institute, New York, 1992.

Air Pollution Primer, National Tuberculosis and Respiratory Disease Association, New York, 1971.

Chow, W. and Connor, K.K., *Managing Hazardous Air Pollutants: State of the Art*, Lewis Publishers, Boca Raton, FL, 1993.

Godish, T. *Air Quality (Second Edition)*, Lewis Publishers, Chelsea, MI, 1991.

Urban Air Pollution in Megacities of the World, World Health Organization/United Nations Environment Programme, Blackwell Publishers, Oxford, 1992.

Williamson, S.J. *Fundamentals of Air Pollution*, Addison Wesley Publishing Co., Reading, MA, 1973.

Periodicals and Journals

Atmospheric Environment, Pergamon Press, New York.

Chemical Engineering, McGraw Hill Publishers, New York.

Econ: The Environmental Magazine for Real Property Hazards, PTN Publishing Inc., Melville, NY.

Environmental Waste Management, International Association of Environmental Managers, Kutztown, PA.

Environmental Protection, Stephens Publishing Corp., Waco, TX.

Environmental Science and Technology, American Chemical Society, New York.

Hazmat World, Tower-Borner Publishing, Inc., Glen Ellyn, IL.

Journal of the Air and Waste Management Association, Pittsburgh, PA.

Journal of the American Meteorological Association, Boston, MA.

Pollution Engineering, Cahners Publishing Company, Newton, MA.

Waste Business West, published by CHMM Inc., Louistown, NY.

Waste Age, National Solid Waste Management Association, Washington, DC.

2. EFFECTS OF AIR POLLUTION

Air Toxics Hot Spots Program Risk Assessment Guidelines, prepared by the AB2588 Risk Assessment Committee of CAPCOA, and the California State Air Resources Board and the California Department of Health Services, January 1991.

Air Toxics Hot Spots Program Facility Prioritization Guidelines, the AB2588 Risk Assessment Committee of the California Air Pollution Control Officers Association (CAPCOA), March 1990.

Analysis of Ambient Data from Potential Toxics Hot Spots in the South Coast Air Basin, Office of Planning and Analysis, South Coast (Calif.) Air Quality Management District, September 1988.

Bolton, J.G., P.F. Morrison, and K.A. Solomon. *Risk Cost Assessment Methodology for Toxic Pollutants from Fossil Fuel Power Plants*, published by the Rand Corporation for the Electric Power Research Institute, Inc., Santa Monica, CA, June 1983.

Hallenbeck, W.H., and K.M. Cunningham. *Quantitative Risk Assessment for Environmental and Occupational Health*, Lewis Publishers, Chelsea, MI, 1986.

Identifying and Regulating Carcinogens, Office of Technology Assessment Task Force (U.S. Congress), Lewis Publishers, Chelsea, MI, 1988.

Kamrin, M. A. *Toxicology: A Primer on Toxicology Principles and Applications*, Lewis Publishers, Chelsea, MI, 1988.

Kopfler, F.C., and G.F. Craun. *Environmental Epidemiology*, Lewis Publishers, Chelsea, MI, 1986.

Lioy, P.J., and J.M. Daisey. *Toxic Air Pollution*, Lewis Publishers, Chelsea, MI, 1987.

Manahan, S.E. *Toxicological Chemistry*, Lewis Publishers, Chelsea, MI, 1989.

Multimedia Health Risk Assessment Input Parameters Guidance Document, prepared for South Coast AQMD by Clement Associates, Fairfax, VA, February 1988.

NATICH Database report on State, Local and EPA Air Toxic Activities, National Air Toxics Information Clearinghouse, U.S. EPA Office of Air Planning and Standards, North Carolina, September 1992.

Public Health Assessment Guidance Manual/ATSDR (Agency for Toxic Substances and Disease Registry), Lewis Publishers, Chelsea, MI, 1992.

Sittig, M. *Handbook of Toxic and Hazardous Chemicals*, Noyes Publications, Park Ridge, NJ, 1981.

Technical Guidance Document for the Emissions Inventory Criteria and Guidelines Regulation for AB2588, California Air Resources Board Technical Support Division, August 1989.

The Magnitude of Ambient Air Toxic Impacts from Existing Sources in the South Coast Air Basin, 1987 Air Quality Management Plan Revision Working Paper No. 3, June 1987, South Coast (Calif.) Air Quality Management District.

3. AIR QUALITY STANDARDS AND MONITORING

Lodge, J.P. *Methods of Air Sampling and Analysis, Third Edition*, Lewis Publishers, Chelsea, MI, 1989.

National Ambient Air Quality Standards, July 1987, U.S. Environmental Protection Agency.

National Air Quality and Emissions Trends Report, 1991, U.S. EPA, Office of Air Quality Planning and Standards, December 1992.

4. SOURCES AND MEASUREMENT METHODOLOGIES

Air Quality Handbook for Preparing Environmental Impact Reports, April 1987, South Coast Air Quality Management District.

National Ambient Volatile Organic Compounds (VOC's) Database Update, prepared by Nero and Associates, Inc., Portland, OR, for the U.S. EPA Office of Research and Development, Atmospheric Sciences Research Laboratory, February 1988.

National Air Quality and Emissions Trend Report, 1991, U.S. EPA, Office of Air Quality Planning and Standards, Document No. EPA-450/4-92/001, December 1992.

SARA 313 Emissions—Air, Water, Underground Injection, Land, POTW's, and Off-Site Releases, U.S. EPA, January 1992.

Steam/Its Generation and Use, The Babcock and Wilcox Company, New York, 1978.

5. METEOROLOGY, DISPERSION, AND MODELING

Guideline on Air Quality Models (Revised) and Supplements, Office of Air Quality Planning and Standards, U.S. EPA, North Carolina, July 1986, and September 1990.

Hanna, S.R., and P.J. Drivas. *Guidelines for the Use of Vapor Cloud Dispersion Models*, Center for Chemical Process Safety of the American Institute of Chemical Engineers, New York, 1987.

Hanna, S.R., D.G. Strimaitis, and J.C. Chang. *Hazard Response Modeling Uncertainty (A Quantitative Method), Volume II: Evaluation of Commonly Used Hazardous Gas Dispersion Models*, prepared by Sigma Research Corporation, Westford, MA, September 1991, for American Petroleum Institute and the Air Force Engineering and Services Center, Tyndall Air Force Base, Florida.

"Introduction to Meteorology," N.E. Bowne and J.E. Yocom, *Chemical Engineering*, a McGraw Hill Publication, July 30, 1979.

"Ozone Air Quality Models: A Critical Review," J.H. Seinfeld, *Journal of the Air Pollution Control Association*, 38(5): May 1988.

User's Network for Applied Modeling of Air Pollution (UNAMAP), Regulatory and Air Quality Simulation Models, U.S. EPA, Office of Research and Development.

Volume 7, *Air Modeling*, Papers from the 84th Annual Meeting and Exhibition, Air and Waste Management Association, Pittsburgh, PA, 1991.

Zannetti, T. *Air Pollution Modeling*, Van Nostrand Reinhold, New York, 1990.

6. STATIONARY SOURCE CONTROL APPROACHES

Air Pollution Engineering Manual (2nd Edition), J.A. Danielson, Ed., AP-40, U.S. Environmental Protection Agency, Office of Air Planning and Standards, May 1973.

Chemical and Process Technology Encyclopedia, D.M. Considine, Ed., McGraw Hill Book Company, New York, 1974.

Handbook of Control Technologies for Hazardous Air Pollutants, U.S. EPA Office of Research and Development, EPA/625/6–91/014, June, 1991.

Hazardous Assessment and Control Technology in Semi-Conductor Manufacturing, prepared by the American Conference of Governmental Industrial Hygienists, Lewis Publishers, Chelsea, MI, 1989.

Hesketh, H.E. *Fine Particles in Gaseous Media*, Lewis Publishers, Chelsea, MI, 1986.

Methane Generation and Recovery from Land Fills, prepared by Emcon Associates, San Jose, CA, for Consolidated Concrete Limited, Ann Arbor Science Publishers, Inc., Ann Arbor, MI, 1980.

Vatavuk, W.M. *Estimating Costs of Air Pollution Control*, Lewis Publishers, Chelsea, MI, 1990.

Workshop on Hazardous and Toxic Air Pollutant Control Technologies and Permitting Issues, Proceedings, U.S. Environmental Protection Agency, Control Technology Center, Cincinnati, OH, presented at San Francisco, CA, April 1988.

7. MOBILE SOURCE CONTROL APPROACHES

Combustion-Generated Air Pollution, Starkman, E.S., Ed., Plenum Press, New York, 1971.

Griffin, R.G. *Principles of Hazardous Materials Management*, Lewis Publishers, Chelsea, MI, 1988.

Handbook of Air Pollution Technology, S. Calvert and H.M. Englund, Eds., John Wiley & Sons, New York, 1984.

Pitts, J.N., Jr. "Atmospheric Pollution in the 1990s: Ozone, Acids, Toxics, and the Greenhouse Effect" (syllabus), Chemistry X471, UCI Education, Winter 1993.

Standard Handbook of Engineering Calculations, Hicks, T.G., Ed., McGraw Hill Book Company, New York, 1972.

8. GLOBAL CONCERNS

Acidic Deposition: Sulphur and Nitrogen Oxides, A.H. Legge and S.V. Krupa, Eds., The Alberta Government Industry Acid Deposition Research Program, Lewis Publishers, Chelsea, MI, 1990.

CRC Handbook of Chemistry and Physics, D.R. Lide, Editor in Chief, 77th Edition, CRC Press, Boca Raton, FL.

Gore, A. *Earth in the Balance: Ecology and the Human Spirit*, Houghton Mifflin Company, New York, 1992.

Idso, S.B. *Carbon Dioxide and Global Change: Earth in Transition*, IBR Press, a Division of the Institute for Biospheric Research Inc., Tempe, AZ, 1989.

Jastrow, R., W. Nierenberg, and F. Seitz. *Scientific Perspectives on the Greenhouse Problem*, The Marshall Press, Jameson Books, Ottawa, IL, 1990.

Ray, D.L., and L. Guzzo. *Trashing the Planet*, Regnery Gateway, Publishers, Washington, DC, 1990.

9. AIR QUALITY LAWS AND REGULATIONS

"Putting the Pieces Together: The Clean Air Act Puzzle," *Waste Age*, March 1991.

Public Law 101-549, *The Clean Air Act Amendments of 1990*.

The Clean Air Act Amendments of 1990 Seminar, Jones, Day, Reavis and Pogue Seminar, Los Angeles, CA, January 1991.

The Clean Air Act Amendments of 1990, Detailed Summary of Titles, U.S. Environmental Protection Agency, Office of Air and Radiation, November 30, 1990.

The Clean Air Act Amendments of 1990, U.S. EPA Seminar, January 1991.

10. MANAGEMENT, TRENDS, AND INDOOR
AIR QUALITY

"Building Consensus Among Business Interests, Environmental Organizations and Regulatory Agencies to Effectively Link Clean Air and Transportation Policies," prepared by the Sequoia Group, the Public Policy Program of UCLA Extension, July, 1992.

Cherry, K.F. *Asbestos: Engineering, Management and Control*, Lewis Publishers, Chelsea, MI, 1988.

Final Air Quality Management Plan, 1991 Revision, South Coast (Calif.) Air Quality Management District, Diamond Bar, 1991.

Godish, T. *Indoor Air Pollution Control*, Lewis Publishers, Chelsea, MI, 1989.

Indoor Air and Human Health, R.B. Gammage, S.V. Kaye, and V.A. Jacobs, Eds., Lewis Publishers, Chelsea, MI, 1985.

GLOSSARY
OF ACRONYMS

AAQ	ambient air quality
ACGIH	American Conference of Governmental Industrial Hygienists
AEL	acceptable exposure level
AQMP	air quality management plan
ARARs	applicable, relevant and appropriate regulations
AVOR	average vehicle occupancy rate
BACM	best available control measure
BACT	best available control technology
BIF	boilers and industrial furnaces
BOF	basic oxygen furnace
BOOS	burners out of service
BRI	building-related illness
BROCS	building related occupant complaint syndrome
BTU	British thermal unit
CAAA	Clean Air Act Amendment
CEM	continuous emission monitors
CERCLA	Comprehensive Emergency Response, Compensation and Liability Act
CFC	chlorofluorocarbon
CO	carbon monoxide
COH	coefficient of haze
CONCAWE	Conservation of Clean Air and Water, Western Europe
CTGs	Control Technique Guidelines
DES	diethyl stilbestrol
DRE	destruction and removal efficiency
EKMA	Empirical Kinetic Modeling Approach
EPA	Environmental Protection Agency
EPCRA	Emergency Planning and Community Right to Know Act
ESP	electrostatic precipitator
ETS	environmental tobacco smoke

FIFRA	Federal Insecticide, Fungicide and Rodenticide Act
FIP	federal implementation plan
GACT	generally available control technology
GCM	general circulation model
HAP	hazardous air pollutant
HCFC	hydrochlorofluorocarbon
HDT	heavy duty truck
HFC	hydrofluorocarbon
ICE	internal combustion engine
ICRP	International Council on Radiation Protection
IDLH	immediately dangerous to life and health
I/M	inspection and maintenance
LAER	lowest achievable emission rate
LDT	light duty truck
LEV	low-emitting vehicle
LFL	lower flammability limit
LPG	liquefied petroleum gas
MACT	maximum available [or *achievable*] control technology
MMH	maximum mixing height
MSA	metropolitan statistical area
MW	megawatt
NAAQS	national ambient air quality standard
NAMS	National Air Monitoring Station
NAPAP	National Acid Precipitation Assessment Program
NESHAP	national emission standards for hazardous air pollutant
NIOSH	National Institute for Occupational Safety and Health
NOAEL	No Observed Adverse Effect Level
NOEL	No Observed Effect Level
NSPS	new source performance standard
NSR	new source review
OCS	outer continental shelf
OEL	occupational exposure level
OSHA	Occupational Safety and Health Administration
PAN	peroxyacyl nitrate
PCB	polychlorinated biphenyl
PEL	permissible exposure level
PIC	product of incomplete combustion
PM	particulate matter
POHC	principal organic hazardous constituent
POTWs	publicly-owned treatment works
PSD	prevention of significant deterioration
PSI	pollutant standards index

psi	pounds per square inch
P-V	pressure-volume
RACT	reasonably available control technology
RCRA	Resource Conservation and Recovery Act
RFP	reasonable further progress
RICE	reciprocating internal combustion engine
RVP	Reid Vapor Pressure
SARA	Superfund Amendments and Reauthorization Act
SCR	selective catalytic reduction
SDCF	standard dry cubic foot
SI	spark ignited
SICs	Standard Industrial Classifications
SIP	state implementation plan
SLAMS	state and local air monitoring station
SNCR	selective noncatalytic reduction
SPM	special purpose monitor
STEL	short-term exposure level
TCA	trichloroethane
TCE	trichloroethylene
TDI	toluene diisocyanate
TLV	threshold limit value
TSCA	Toxic Substances Control Act
TSDFs	treatment storage and disposal facilities
TWA	time-weighted average
TPY	ton per year
ULEV	ultra-low emission vehicle
URF	unit risk factor
UV	ultraviolet
VMT	vehicle miles traveled
WHO	World Health Organization
ZEV	zero emission vehicle

INDEX